AIRPORT SPOTTING GUIDES

UNITED KINGDOM & IRELAND

2ND EDITION

DestinWorld publishing

Second Edition 2022

ISBN 978-1-7398194-3-9

British Library Cataloguing-in-Publication Data

A catalogue record for this book is available from the British Library.

Published by Destinworld Publishing Ltd.

www.destinworld.com

Printed in the United Kingdom.

FSC
www.fsc.org

Cover design by John Wright

Table of Contents

England

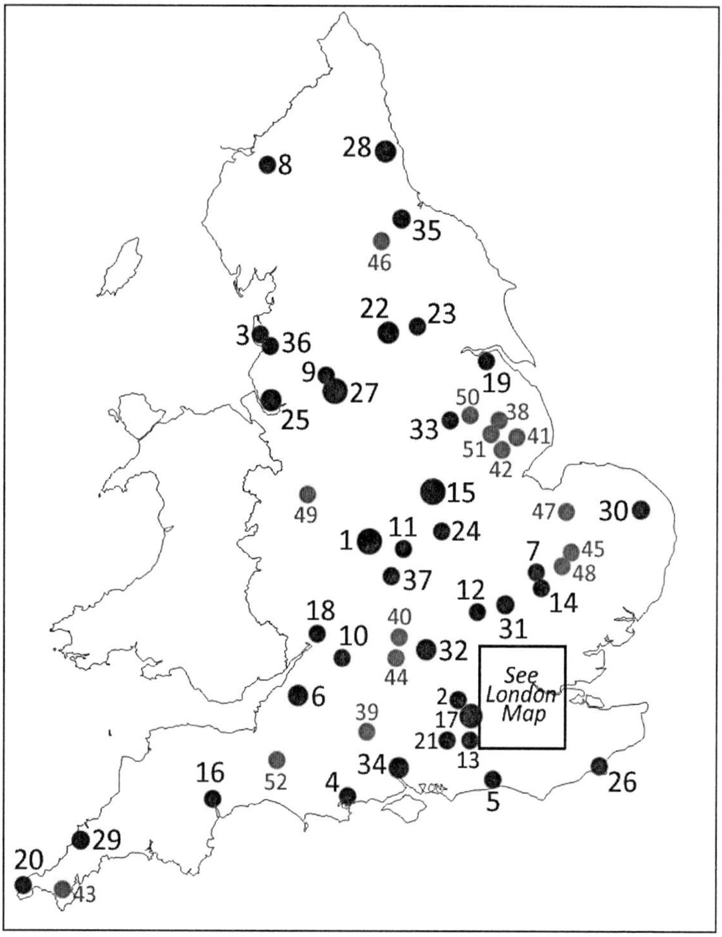
28
8
35
46
22
23
3
36
9
27
19
50
38
25
33
51
41
42
15
49
47
30
11
24
1
7
45
48
37
12
14
18
40
31
10
32
44
See London Map
6
2
39
17
21
13
16
52
34
4
5
26
29
20
43

1. Birmingham
2. Blackbushe
3. Blackpool
4. Bournemouth
5. Brighton City Shoreham
6. Bristol
7. Cambridge
8. Carlisle Lake District
9. City Airport & Manchester Heliport
10. Cotswold Airport
11. Coventry
12. Cranfield
13. Dunsfold
14. Duxford
15. East Midlands
16. Exeter
17. Farnborough
18. Gloucestershire
19. Humberside
20. Land's End
21. Lasham
22. Leeds Bradford
23. Leeds East
24. Leicester
25. Liverpool John Lennon
26. Lydd
27. Manchester
28. Newcastle
29. Newquay
30. Norwich
31. Old Warden
32. Oxford
33. Retford Gamston
34. Southampton
35. Teesside
36. Warton
37. Wellesbourne Mountford
38. RAF Barkston Heath
39. MoD Boscombe Down
40. RAF Brize Norton
41. RAF Coningsby
42. RAF Cranwell
43. RNAS Culdrose
44. RAF Fairford
45. RAF Lakenheath
46. RAF Leeming
47. RAF Marham
48. RAF Mildenhall
49. RAF Shawbury
50. RAF Scampton
51. RAF Waddington
52. RNAS Yeovilton

Birmingham

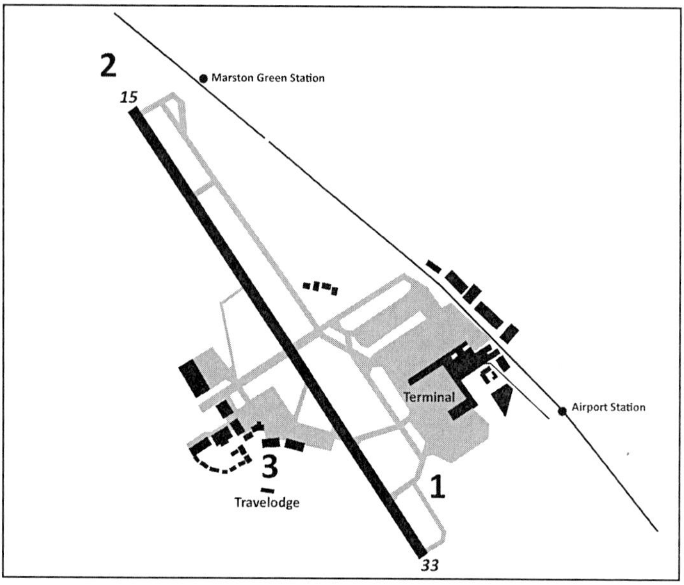

Runways

15/33 10,013ft / 3,052m

Hub For: Flybe, Jet2, Ryanair, TUI Airways

Birmingham is a relatively busy and important international gateway. It doesn't offer much more than can be found at Manchester or London's main airports, but it is worth a visit when in the area. Aside from the heavy presence of Jet2, Ryanair, and TUI, other carriers of note include Emirates, Pakistan International, Turkish, and Wizz Air. Europe's full service carriers are also well represented, including Aer Lingus, Air France, Brussels Airlines, KLM, Lufthansa, Swiss and Scandinavian Airlines.

Birmingham's modern terminal occupies the south western portion of the site, with two piers. Across the runway is the former Elmdon Airport and its historic terminal. Today biz jets and cargo aircraft park here, alongside a maintenance hangar which sometimes sees unusual visitors.

There are a few spotting locations around the airport, but none at the terminal itself. Spotters often enjoy taking video and pictures of aircraft struggling with the prevalent crosswinds at Birmingham.

SPOTTING LOCATIONS

1. Long Stay Car Park

The airport provides an official viewing area in the Long Stay Car Park alongside the runway 33 threshold and taxiway. There are benches and a good view of aircraft movements through the fence, and shortly before touchdown. Follow signs from the terminal.

2. Runway 15

At the northern end of the airport this spot is good for approach and crosswind photographs. A path here runs across the approach path. It is best accessed from Hazeldene Road (postcode B33 0QB), off The Radleys. If you're heading from Birmingham city centre, follow the A45 and then turn left onto Sheaf Lane in Sheldon. This turns into Church Road. After a mile turn right onto The Radleys, and right again opposite the car garage. Alternatively, get the train to Marston Green station (one stop from the airport/NEC) and walk along the path to the viewing area.

3. Elmdon Multi-Storey

The multi-storey car park next to the old Elmdon terminal on the western side of the airport has good views from the top floors. You can see and photograph cargo, executive and general aviation aircraft here, and have good views of the runway. Photography better on the afternoon.

HOTELS

Travelodge Birmingham Airport

Terminal Road, Birmingham B26 3QW |
+44 871 984 6483 | *www.travelodge.co.uk*

Located on the old terminal side of the airport, rooms facing the airport on the top two floors have great views across the runway and to the passenger terminal. You can't see the executive and cargo ramp, but will see all movements.

Public Transport: Birmingham Airport has its own railway station, shared with the NEC exhibition centre. Trains run from Birmingham itself, as well as cities such as London and Manchester. There is an Air-Rail link between the station and the terminal. Buses also serve the airport.

Blackbushe

Runways

07/25 4,380ft / 1,335m

It's hard to imagine today that, after World War II, Blackbushe became one of London's two primary civil airports handling flights from around the world. This ended in the early 1960s when the airport reopened as a general aviation facility. Today it is fairly busy with light aircraft and sees a good mix of executive aircraft visiting. It's close enough to Farnborough for a quick visit.

Blackbushe has one runway, with an area of grass and concrete parking at the south east corner of the site, next to the small handling facilities and flying clubs.

Most aircraft can be seen from the fence next to the car park outside the control tower. A short walk along the fence line next to the A30 road should yield anything missing.

Public Transport: Blackbushe is around 5 miles from Farnborough, and 23 miles from Heathrow Airport. There are no public transport links to the airport, but railway stations in Farnborough and Blackwater are fairly close.

Blackpool

Runways

10/28 6,132ft / 1,869m
13/31 3,294ft / 1,004m

Blackpool Airport has had mixed fortunes in recent years, going from a busy regional airport with based aircraft of Jet2, to closure in 2014. Thankfully the airport was reopened in April 2015 and has since undergone some redevelopment. There are no passenger services any more. Helicopter services to Irish Sea oil rigs and general aviation make up the majority of movements.

Blackpool has two active runways, with all service buildings running along Squires Gate Lane which runs from the nearby sea front. It's possible to see parked aircraft from gaps between the hangars, and also from the fence line behind the Morrissons supermarket a little further along. Some operators are known to allow people to visit the hangars during quieter periods.

Public Transport: Buses #11 and #68 stop nearby the airport, linking it with the town centre. Bus #7 also runs along Squire's Gate Lane.

Bournemouth

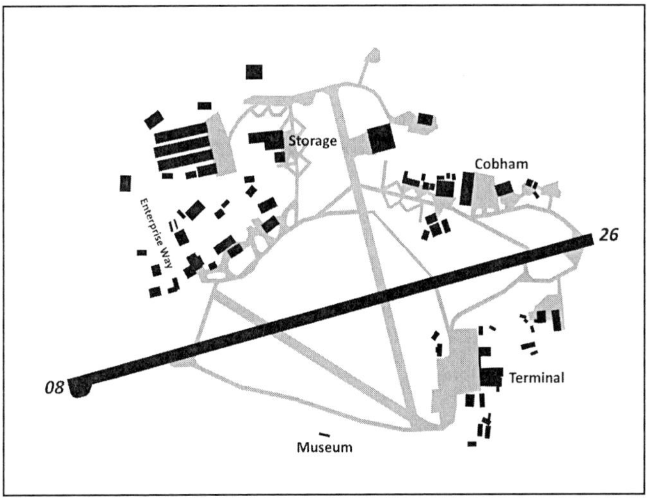

Runways

08/26 7,451ft / 2,271m

Hub For: Ryanair, TUI Airways

Bournemouth is an interesting airport in the South West of the country. For years it was a manufacturing plant for airliners such as the BAC One-Eleven and Vickers Viscount. Today a good amount of maintenance work still goes on, so it's usual to see some airliners and other aircraft parked outside the hangars on the north side.

Airliner traffic has been up and down. Most flights are with Ryanair's Boeing 737s as one of the airline's smaller hubs, with TUI also flying a good selection of seasonal and year-round routes.

The airport is split into three distinct segments, separated by the main east-west runway and former cross runways. At the south east corner is the passenger terminal, apron and car parks. It is possible to follow the road here until you reach the perimeter fence and view aircraft parked in the Draken areas and around the terminal, but don't loiter.

In the north east are flying clubs, air ambulance and a large facility operated by Draken Europe. A drive along Mountbatten Drive (postcode BH23 6NE) will offer some glimpses through the fence.

The north west corner comprises many hangars and buildings associated with maintenance, storage, business aircraft, a Pilatus PC-12 centre, and non-aviation businesses. Enterprise Way (postcode BH23 6BS) is the main road running through this business park, and you are free to pass through the security post. Work your way past the various small roads running to different areas and spot what you can through the fence. At the end you'll come to the area used for airliner storage.

Public Transport: Yellow Bus #737 links Bournemouth city centre with the airport passenger terminal. It also stops outside the museum (see below) and Enterprise Way in the north west section of the airport.

The **Bournemouth Aviation Museum** occupies a site off the B3073 on the southern perimeter. Open daily from 10am except Christmas bank holidays, with a selection of civil and military aircraft and cockpits. A double-decker bus here offers views across the airport from above the fence line. www.bamhurn.org

Brighton City Airport

Runways

02/20 3,399ft / 1,036m
02G/20G 2,297ft / 700m
07G/25G 2,877ft / 877m
13G/31G 1,339ft / 408m

Known by most as Shoreham Airport, Brighton City is situated equidistant between Brighton and Worthing on a plain alongside the Rive Adur and the English Channel. It has one concrete runway and two grass runways, and is known as a centre for flight training.

The airport does not have any passenger airline service, but has a lovely Art Deco terminal building in front of the aircraft parking apron and amongst the many small hangars running along the access road.

The airport is used mostly by light aircraft, pleasure flights, helicopters and vintage aircraft. Follow Cecil Pashley Way past the terminal and you'll find a number of parking places along the fence line to watch movements from.

Public Transport: The nearest buses pass along the A259, and Shoreham railway station is a 20 minute walk from the airport.

Bristol

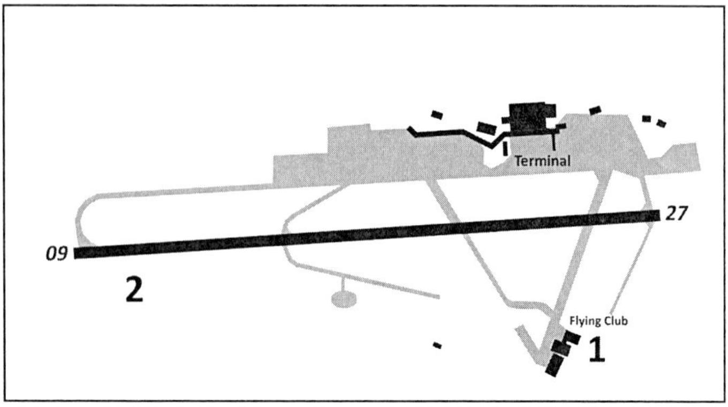

Runways

09/27 6,598ft / 2,011m

Hub For: easyJet, Jet2, Ryanair, TUI Airways

Bristol Airport is at Lulsgate, a few miles south of the city in South West England. It has a single runway and modern terminal and handled almost 9 million passengers in 2019. Passenger airliner parking aprons are found on the north side of the runway, which is in an east-west orientation. To the south, light aircraft, executive aircraft and based flying clubs use parking areas on the former cross runways.

Bristol is a hub for a few low-cost and leisure airlines. Other carriers include Aer Lingus Regional Aurigny, Blue Islands, KLM Cityhopper, Loganair, Lufthansa, and Wizz Air. The summer season sees many more charters and seasonal flights.

A dedicated spotting area exists at the flying club on the south side of the airport, known as the Fight Lounge (1). Following the perimeter road from the terminal to the 'Silver Zone Car Park' and then find the club signposted. Views from the café here look across the runway to the terminal area and are good for watching light aircraft and bizjet movements.

Another popular location is Winters Lane. From the airport passenger terminal entrance, drive north on the A38 and take the first left along Downside Road. After a mile turn left along Cooks

Bridle Path, signposted Tall Pines Golf Club. Pass the golf club and you'll reach the airport perimeter. Winters Lane follows this around the end of runway 09 to the south side where a car park offers views through the fence (2). Photography is difficult without a step ladder.

Public Transport: None of the spotting locations are accessible by public transport but can be walked from the passenger terminal. Regular buses link the terminal with Bristol city centre and train station.

Cambridge

Runways

05/23 6,446ft / 1,965m

A small airport on the eastern edge of the city. It has one concrete and two grass runways, with all of the aprons, hangars and other buildings on the north side of the site. Cambridge doesn't have any regular airline services, but does see plenty of airliner movements coming in for maintenance at Marshalls Aerospace, including military transport aircraft. Executive aircraft are also common.

Airport Way (head for Church Road, CB1 9AZ) runs along part of the eastern perimeter, with places to pull over and look across the airfield through the crash gates (don't obstruct).

At the southern end of the airport, a small residential street called Hatherdene Close (postcode CB1 3HQ) off Coldhams Lane has views across the runway to the Marshalls hangars.

Finally, at the northern part of the airfield you can park up near the flying club hangars and see some of the parked aircraft nearby.

Cambridge Airport is due to close by 2030 for redevelopment of the site.

Public Transport: Buses #10, 11 and 12 link the city centre with the airport and pass Airport Way. Bus #17 passes Hatherdene Close.

Carlisle Lake District

Runways

01/19 3,077ft / 938m
07/25 6,027ft / 1,837m

A small airport which has never developed any significant traffic, despite a new terminal opening in 2019 and a brief spell of scheduled services operated by Loganair prior to the Covid-19 pandemic. Carlisle has two runways and a small central terminal and hangar area. It is mostly used by light aircraft for training, along with the occasional biz jet and military helicopter visitors. You can see all parked aircraft from the access road and car park on the north side (not the passenger terminal).

The access road passes the Solway Aviation Museum next to the airport. It has a collection of historic military jet aircraft including a complete Vulcan bomber. Open weekends and public holidays from April to October www.solway-aviation-museum.co.uk

Public Transport: Bus #BR1 passes the airport entrance, linking it to central Carlisle.

City Airport & Manchester Heliport

Runways

08R/26L 2,051ft / 625m
08L/26R 2,103ft / 641m
02/20 1,749ft / 533m
14/32 1,306ft / 398m

Acting as the general aviation gateway to the city of Manchester, this is a very historic airfield and an interesting place to visit.

Previously known as Barton Airfield, City Airport is located just off the M60 and M62 motorways, near the Trafford Centre. It's around a ten minute drive from Manchester Airport.

The airfield is home to lots of light aircraft, and is also a busy heliport, with based Police unit.

There's a nice café at the heart of the airfield with views, and you can wander around to see parked aircraft. The historic control tower also has a balcony you can climb up to.

Cotswold Airport

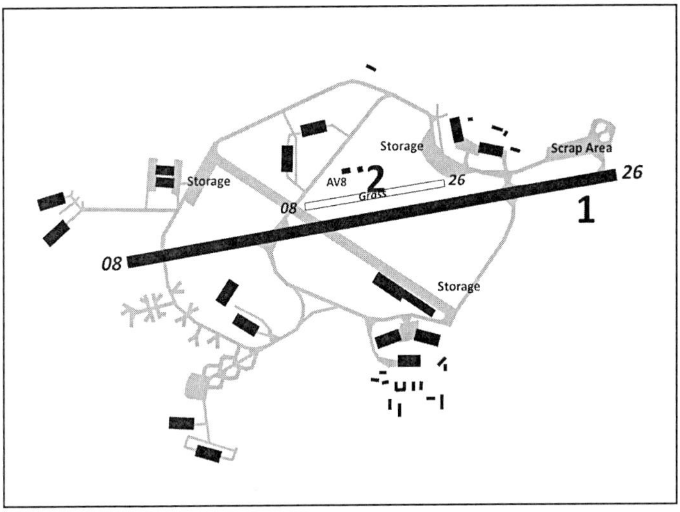

Runways

08/26 6,561ft / 2,009m
08G/26G 1,837ft / 560m

Located at Kemble in Gloucestershire, this former wartime airfield is now a busy general aviation airfield and a huge draw for enthusiasts thanks to the large numbers of airliners which are ever present awaiting parting-out and scrapping. Aircraft from all over the world, particularly Airbus and Boeing airliners, arrive regularly and park on the various ramps and taxiways around the airfield before they work their way to the scrapping area in the north-east corner.

Also on site are Chevron and Lufthansa Technik which provide aircraft maintenance and training, with a few of their aircraft sometimes visible outside their hangars.

Resident preserved and training aircraft include a Royal Air Force Bristol Britannia, VFW-614, a Folland Gnat, and a couple of Hawker Hunter jets.

The A429 leading out of Kemble village passes the fence line where some store airliners can be read off (1). You can park in the layby, and look across to the scrapping compound.

Head back to the village follow signs for the airport and work your way from the entrance to the AV8 Restaurant and control tower, where you can park and will see other stored aircraft from different vantage points (2). Former British Airways Boeing 747 G-CIVB (in the Negus retro livery) is preserved as an events venue.

Public Transport: There is a railway station in Kemble village, but no public transport to the airport itself so you would need to walk around 3 miles.

Coventry

Runways

05/23 6,586ft / 2,008m

Another airport likely to close in the near future, with a Gigafactory expected to be built on the runway.

Despite many attempts, Coventry has never really made it as an airport for passenger flights. However, for many years it served some interest to the enthusiast as a home for preserved, vintage and stored aircraft. These have sadly now all but disappeared.

Most activity takes place on the western extreme of the airport, where occasional cargo flights park. The northern portion of the airport is home to flying clubs and the Midland Air Museum, which has a nice collection of vintage civil and military aircraft. The museum opens daily at 10am www.midlandairmuseum.co.uk

Rowley Road runs along the northern boundary of the airport, passing the museum, with some views from the flying club car park. Follow the road west until you come to Coventry Road. Turn left along it, passing through Baginton village. Pass the end of the runway and turn left onto a small road running parallel to the airport fence. From here you can see across the airfield and runway, and see a retired Lockheed L-188 Electra.

Cranfield

Runways

03/21 5,902ft / 1,799m

Cranfield is an interesting airport to visit. Located near Bedford, it was a former World War II base and today acts as a location for general aviation, business flights, and research and development activities.

The airport is not particularly busy, but you might like to call in to log some of the long-term residents, including a Boeing 737-400, and various British Aerospace types, like Jetstreams and 146/Avro RJ airliners

In the future, Cranfield is expected to become the new base for Marhsall Aerospace when it leaves Cambridge Airport.

From Cranfield village (on the eastern side of the airport), you can either head south on High Street. At the roundabout, turn right onto Stillitters Farm Road. Moving along here gives you views of the south western part of the airfield.

Alternately, from central Cranfield village, head north along Merchant Lane, which will pass the northern end of the airfield.

You can explore the roads around the Cranfield University and technical site, but views are fleeting.

Dunsfold

Runway

07/25 5,496ft / 1,675m

A former military airfield in Surrey, to the west of London Gatwick. Movements are fairly limited, but visitors will be interested in the three Boeing 747s preserved on site (including one of the special livery British Airways machines), plus a Vickers VC-10 and some other airframes. These are mostly used for filming duties, but are occasionally open to visitors.

A drive around the technical site and hangars along the northern side of the airfield affords views across to the airliners on the southern side of the runway, albeit at a distance.

Duxford

Runways

06L/24R 2,887ft / 880m
06R/24L 4,931ft / 1,503m

One of the principal aviation attractions in the UK. Duxford is a former wartime airfield of historic significance. Today it is home to the Imperial War Museum's main aviation collection, including that of the American Air Museum and the British Airlines Collection. It can take hours to work your way around the incredible collection of preserved aircraft from all major conflicts and civil eras, including a Concorde prototype.

Duxford is an active airfield for light aircraft and pleasure flights, and throughout the year it hosts some of the best air displays in the country, often with aircraft not seen at other shows.

Open daily from 10am (except Christmas bank holidays). Visit www.iwm.org.uk for more information.

Public Transport: Bus #132 from Cambridge runs to Duxford on Sundays. There is a Park and Ride car park at Trumpington with free shuttle bus.

East Midlands

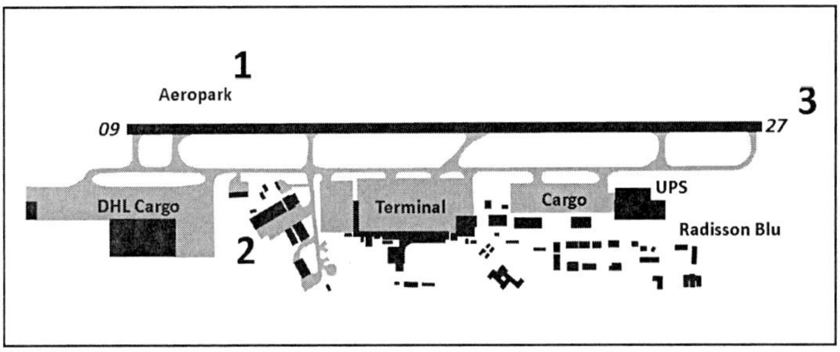

Runways

09/27 9,491ft / 2,893m

Hub For: DHL Aviation, Jet2, Ryanair, TUI Airways, RVL Aviation, UPS Airlines

East Midlands is probably the UK's most notable cargo airport. It successfully transitioned from a quiet regional airport by developing as a hub for cargo airlines such as DHL and UPS. These now serve the airport daily alongside other carriers such as Cargolux, FedEx Express, West Atlantic, Star Air, Swiftair and Lufthansa Cargo. This is not to belittle the airport as a passenger base, with airlines such as Ryanair, Jet2, TUI all offering a decent number of services.

Cargo flights operate mostly at night between 2100-0200 from Sunday to Friday, which is a little unsociable for spotting. During the daytime you will only likely see a few DHL aircraft parked up, although there are usually more present on Saturdays. UPS and FedEx aircraft park on the eastern cargo ramp, whilst all others use the large dedicated DHL ramp at the western extreme of the airport.

Often unusual aircraft are visiting the paint hangars near spotting location 2 (see below).

Beware that the roads around the entire terminal and hangars area is a red route, which means cars are not permitted to stop anywhere apart from the car parks. This makes it difficult to log aircraft at the maintenance hangars and general aviation area.

SPOTTING LOCATIONS

1. Castle Donington Crash Gate

Most spotters congregate at the crash gate on the northern side of the airfield as it is an accepted place to spot and offers the best views of all movements. To reach the gate, head to the village of Castle Donington. Close to the Aeropark museum, there's a pub with a small road next to it (Diseworth Road, postcode DE74 2PS). This road leads to the crash gate. You can see the passenger terminal and DHL apron, and all runway movements. It is good for photography through holes in the fence and once aircraft are above the fence line. You can also wander along the path which extends along the airport's northern boundary from here.

2. Hangar Area

From the airport's long stay car parks and access roads near DHL, you can drive to an area among the hangars which will yield any aircraft receiving maintenance, as well as some executive and light aircraft. Don't stop here, but try driving along Anson Road (DE74 2SA) which will yield aircraft at the RVL hangar, some of the DHL ramp.

3. Runway 27 Viewpoint

Near the end of runway 27 is a good place to park up and watch aircraft movements. Photography is sadly into the sun. The northern perimeter track also extends from this crash gate, eventually reaching location 1. Head for Ashby Road postcode DE74 2DJ, reached off the roundabout on the A453 where it meets the A6 at Wilders Way.

East Midlands Aeropark is a nice outdoor museum of aviation exhibits, including a Nimrod, Argosy and Vulcan, and has good views of runway movements. Opens Thursday, Saturday and Sunday in summer, and Sundays only in winter www.eastmidlandsaeropark.org

HOTELS

Radisson Blu East Midlands

Pegasus Business Park, Herald Way, East Midlands Airport DE74 2TZ | +44 1509 670575 | *www.radissonhotels.com*

Many rooms overlook the threshold of runway 27 and parts of the eastern cargo ramp. You can see most movements, and photograph aircraft arriving/departing with a 300mm lens. The hotel is located to the east of the terminal and cargo areas.

Public Transport: East Midlands Airport is linked to Derby, Leicester, Loughborough and Nottingham (and their railway stations) via bus services. Bus #155 from the passenger terminal also stops next to the Aeropark and spotting location 1 in Castle Donington.

Exeter

Runways

08/26 6,811ft / 2,076m

Hub For: TUI Airways

A modest regional airport in South West England serving the local population. It primarily sees regional services by the likes of Aer Lingus Regional, Aurigny, Blue Islands and Loganair. Ryanair and TUI offer holiday routes. There are some cargo services, too.

The passenger terminal and Exeter Aerospace's large hangar and apron occupy the southern boundary of the airport, with the east-west runway alongside. Further north of the runway former taxiways and the disused cross runway are usually home to a few stored airliners awaiting their next assignments.

There are no official viewing areas at Exeter any more. You can get views across the runway and terminal apron from long stay car park P3, but it is through a fence so photography is no good.

Glimpses of anything stored on the north side are possible from the B3174 London Road as it heads north. It's easier if you find somewhere to park and then walk alongside the road.

Public Transport: Bus #56 links Exeter Airport with the city centre throughout the day.

Farnborough

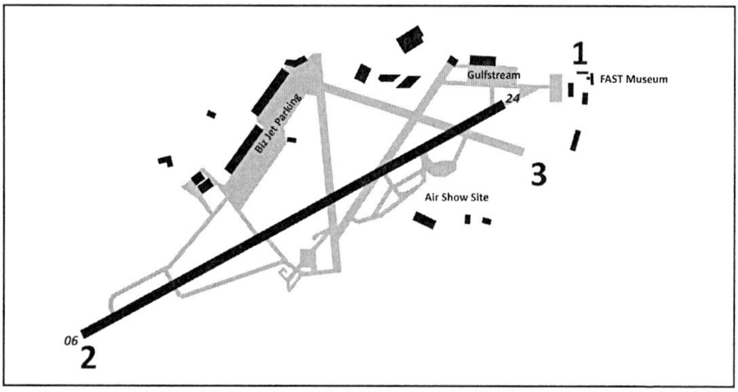

Runways

06/24 8,005ft / 2,440m

Most famous for its biennial air show, Farnborough has been the cradle for aviation development and testing in Britain for a long time. The airport is now a busy hub for biz jets bringing their owners to London and the South East, making it particularly attractive to enthusiasts. Gulfstream Aviation recently opened a large hangar.

The airport site is encroached upon by housing and the nearby town centre. Its main runway separates the two sides to the airport, with the control tower, FBOs and main parking aprons along the northern side. The southern side is reserved for the air show events, with large pavilions and parking aprons designated for visiting aircraft.

SPOTTING LOCATIONS

1. Rae Road/FAST Museum

The cul-de-sac alongside the FAST Museum, Rae Road (postcode GU14 6XE) is more or less in line with runway 24. You can look along its length, and approaching aircraft will pass very low overhead. Walk a little along the main road to get a good vantage point, or read off (with strong binoculars) aircraft parked up on the north side.

2. Eelmore Bridge

At the 06 end of the runway is a distinctive iron bridge off the main A323 Fleet Road. It is technically on Laffan's Road (postcode GU11 2HL) and has a small area for parking. This location has great views of aircraft arriving and departing runway 06, and of departures off 24. You can see aircraft parked on some of the ramps, but not all.

3. Trenchard Way

When the air show is not on, Trenchard Way (postcode GU14 6GU) is a good route to follow (preferably park the car and walk along it) as it offers views across the runway to the various parking aprons on the north side.

Farnborough Air Sciences Trust (FAST) Museum is dedicated to the aviation history of Farnborough and its involvement in the development of aircraft and associated technology. It has some interesting displays and a number of aircraft outside. Open weekends and public holidays from 10am to 4pm. Visit www.airsciences.org.uk

Public Transport: From Farnborough Central Station, you can take bus #42 which passes the FAST Museum and Queen's Gate Road.

Gloucestershire Airport

Runways

04/22 3,241ft / 988m
04G/22G 997ft / 304m
09/27 4,695ft / 1,431m
18/36 2,621ft / 799m

Gloucestershire Airport is one of Britain's most popular airfields for general aviation, and also sees a good number of executive aircraft movements and helicopters. After years following elusive scheduled services, the former wartime training base now has no airline routes.

The Aviator café is probably the best place for views. It serves food and drink and has outdoor seating with views over the airfield, flying clubs and runway 27 threshold, which is great for photographs. You can find it next to the airport entrance, off Bamfurlong Lane (postcode GL51 6SR).

Driving or walking further along the road into the hangar area should yield more parked aircraft outside the different companies and organisations present.

Jet Age Museum

At the north west corner of the airport site, in Meteor Business Park (postcode GL2 9QL) is this small museum with a number of aircraft exhibits on display. Open weekends from 10am to 4pm. See www.jetagemuseum.org

Public Transport: Bus #94 runs from Cheltenham to Gloucester, passing along the airport's northern perimeter and the Jet Age Museum. It is a short walk to The Aviator if you alight at Bamfurlong Industrial Park in Staverton Bridge.

Humberside

Runways

02/20 7,205ft / 2,196m
08/26 2,822ft / 860m

Hub For: Bristow Helicopters, Eastern Airways

Humberside is a small regional airport in North Lincolnshire, between Scunthorpe and Grimsby, and across the river from Hull. It is not busy, but sees daily scheduled services from KLM Cityhopper and Eastern Airways (which maintains its fleet here). Summer charters add to the mix, and there are plenty of helicopter flights to the North Sea oil rigs, and light aircraft movements from local training schools.

Views of anything parked near the passenger terminal or helicopter terminal can be obtained from the car parks and access roads.

A good location for watching arrivals on runway 21, where locals seem to prefer, is a small layby off the A18 along a road signposted for Grasby and Caistor. You can see across the runway to the parking ramps, and a little further along see part of a BAC 1-11 fuselage (G-AVMP) used for fire training near the smaller cross runway. This location is a 15 minute walk from the terminal.

Public Transport: Direct bus services no longer run to the airport, but JustGo operate a passenger request service. The nearest train station is Barnetby, three miles away.

Land's End

Runways

02/20 1,591ft / 485m
07/25 2,221ft / 677m
12/30 1,673ft / 510m
16/34 2,572ft / 784m

Land's End serves the most south-westerly tip of the Great Britain mainland in Cornwall. It is located a few miles west of Penzance, the nearest main town.

While it sees some general aviation activity, the airport's main reason for being is in providing a regular air link to the Isles of Scilly, off the coast. These are provided by Isles of Scilly Skybus, with their DHC-6 Twin Otters and Britten Norman Islanders throughout the day. You can see their aircraft parked on the ramp and around their hangar from the car park and terminal area, with photographic opportunities.

Lasham

Runways

09/27 5,896ft / 1,797m

This small airfield in Hampshire has a long association with aircraft storage and maintenance. For many years it was the home to Dan-Air London's maintenance division.

Today, Lasham is a busy gliding base, with a club on the northern boundary. It's also still a site for airliner maintenance (with some long-term storage and scrapping), particularly with end-of-lease work. As such you'll often see a handful of airliners parked up.

It's not possible to see all aircraft from one location. The best locations to see them are from the crash gate at the northern end Lasham village (head past the Royal Oak pub, postcode GU34 5SL). From here you can see some of the stored airliners (look out for the derelict aircraft in the trees on your left).

Another good location is the café of the Lasham Gliding Society, off The Avenue (postcode GU34 5SS) which runs parallel to the runway. You can see the many gliders and have views across to the hangars and stored airliners.

Leeds Bradford

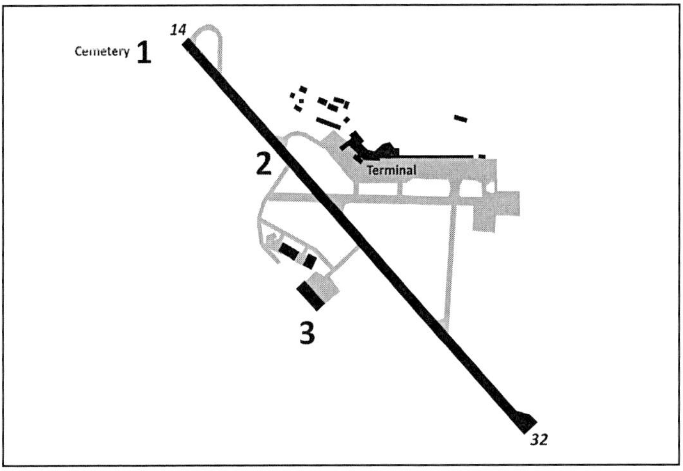

Runways

14/32 7,382ft / 2,250m

Hub For: Jet2, Ryanair

Leeds Bradford is one of the busier airports in northern England, linking the twin cities in Yorkshire. It is Britain's highest airport, at 680ft above sea level, and its exposed position often makes for a windy approach.

The airport has one runway. The passenger terminal is on the north side, whilst to the south are maintenance hangars and the complex use for general aviation and flight training. Spotting opportunities landside are limited around the terminal.

The main operators are Jet2, with a large presence, plus Ryanair, Aer Lingus Regional, KLM and TUI Airways.

SPOTTING LOCATIONS

1. Cemetery

Walking to the north west you'll see the road tunnel which passes underneath the runway on your left. Don't go through the tunnel, but continue over the roundabout, then left at the T-junction. You'll come to a cemetery which is used by local spotters which is great for runway 14 operations.

2. Waste Ground

Alternatively, go through the tunnel and continue to a crossroads. Turn left onto Bayton Lane. A little further on your left is an area of waste ground overlooking parts of the airport and runway. You can't drive onto the area any more, so will need to park elsewhere and walk. The next left also leads to a crash gate and a path along parts of the southern perimeter.

Public Transport: Bus #757 passes under the runway tunnel as it heads to and from the airport. It's possible to walk from the south side of the tunnel to any of the spotting locations in less than 20 minutes.

Leeds East

Runways

06/24 5,610ft / 1,710m
1634 4,806ft / 1,465m

The former RAF Church Fenton is now in civilian hands and operating as Leeds East Airport. It's located near Selby and Tadcaster in North Yorkshire.

While ambitions to attract commercial flights have faltered, the number of general aviation and even executive movements has grown. The roads around the hangars, and the Fenton Flyer café, have some views of parked aircraft and movements.

Leicester Airport

Runways

04/22 1,608ft / 490m
06/24 1,115ft / 340m
10/28 3,068ft / 935m
15/33 1,624ft / 495m
16/34 1,371ft / 418m

This general aviation airfield is a few miles outside of Leicester. It has no commercial services, but can be busy with helicopter training activities. Light aircraft movements are busy on a weekend.

There's a café in the tower with views of the main runway 10/28. The car park also has views.

Liverpool John Lennon

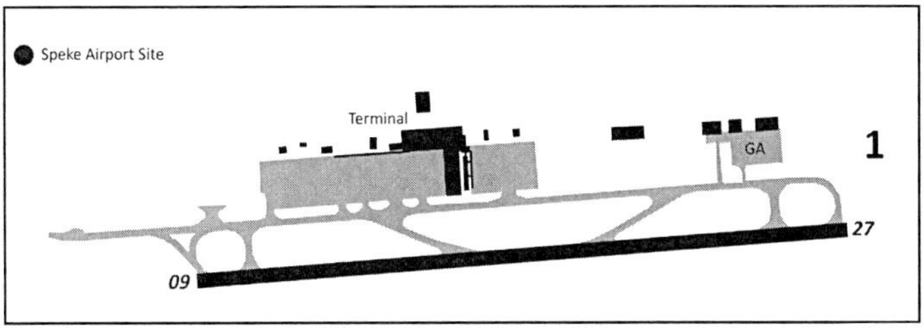

Runways

09/27 7,497ft / 2,285m

Hub For: easyJet, Ryanair

Despite living in the shadow of the much larger Manchester Airport some 30 miles away, Liverpool has still managed to carve a successful operation from its substantial catchment area across the North West and North Wales. Its primary operators are low-cost and leisure carriers, and regional links – particularly across the Irish Sea.

Aside from easyJet and Ryanair, other carriers include Blue Air, Loganair, Lufthansa, and Wizz Air.

Liverpool's modern terminal offers views once through departures, but is fairly restrictive for those without a ticket. For the best location head east from the terminal on Hale Road, then turn right onto Dungeon Lane and right again toward the flying school. There are places to park (1), and you will see aircraft on the GA apron and using runway 27. Photographs are possible (into the sun).

To the north of the present airport on the old Speke Airport site the historic original Art Deco terminal is now a Crowne Plaza hotel, with a small collection of historic airliners being preserved outside including a Bristol Britannia, HS.748 and Jetstream 41.

Public Transport: Bus #89 from the airport passes the end of Dungeon Lane and the GA hangar every 20 minutes.

Lydd

Runway

03/21 4,938ft / 1,505m

Lydd is an airport on the south coast which has a long history as a gateway to France. This was particularly true during the era of car transporter aircraft, which would cross the Channel many times per day with passengers and their vehicles.

Sadly there are no airline links from Lydd at present, and the airfield is only used by general aviation traffic and private flights, plus a HM Coastguard base.

The car park outside the terminal is fine for logging any aircraft on the main apron, while the access road has views through the fence of the northern apron.

Manchester

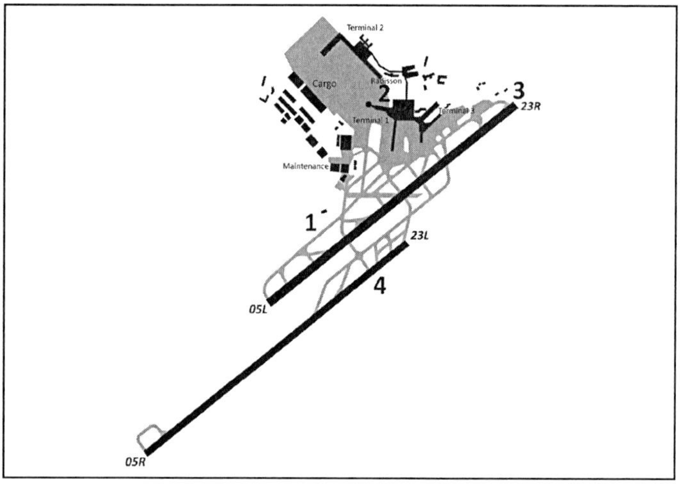

Runways

05L/23R 10,000ft / 3,048m
05R/23L 10,007ft / 3,050m

Hub For: Aer Lingus UK, easyJet, Jet2, Ryanair, TUI Airways, Virgin Atlantic

Manchester is one of the favourite airports in the UK for aviation enthusiasts thanks to its excellent viewing facilities and good mix of movements. The airport can be busy at certain times of the day, particularly in the summer months.

Manchester is one of the UK's busiest airports and has long been popular with both charter and scheduled carriers, and today is a hub for low-cost carriers too. The north of England gateway is the third busiest airport in the country, after Heathrow and Gatwick.

Long haul flights at Manchester have seen increasing presence in recent years. It sees the likes of Air Canada, Air Transat, Cathay Pacific, Biman Bangladesh, Emirates, Ethiopian Airlines, Etihad, Gulf Air, Kuwait Airways, Pakistan International, Qatar Airways, Saudia, and Singapore Airlines. Emirates operates the Airbus A380 here. A number of cargo carriers also provide daily widebody aircraft for added interest, including FedEx Express.

Manchester Airport has three passenger terminals and a cargo terminal, along with an extensive maintenance area comprising large hangars and bays. Terminal 2 is being developed into the main passenger terminal with additional piers. Eventually Terminal 1 will be removed. It has two parallel runways.

SPOTTING LOCATIONS

1. Runway Visitor Park

The Runway Visitor Park was built on the north side of Runway 05L/23R to replace facilities lost when the new runway was built and rooftop terrace closed. As well as a raised area to view aircraft on the runway and taxiway (which has sadly been restricted due to a new building on site), it also has an aviation shop, café, toilets and various preserved airliners – some of which are open to the public. These include a DC-10 section, British Airways Concorde, Trident,

Nimrod and Avro RJX. The park is open daily (except 25/26 December) from 8am till dusk (4pm in winter, 6pm in spring/autumn, 8pm in summer). There is a car parking fee. Those on foot/bike enter free. www.runwayvisitorpark.co.uk

2. Multi-Storey Car Park

For many years the top level of the Short Stay Car Park outside Terminal 1 was one of the main spotting locations. It is still a popular stop off as it offers fantastic views over the cargo and maintenance ramps, as well as parts of Terminal 2, Pier C of Terminal 1, and the runways in the distance. Spotting and photography are still good from the car park's open roof. This is a good spot to log what's not visible from the south side, but parking charges can prohibit extended stays.

3. Airport Pub

Located at the threshold of Runway 23R. The beer garden here backs up to the taxiway and holding point for the runway and is ideal for landing shots. A very pleasant summer afternoon can be held here, with food and refreshments on tap. Car parking charges apply, or it is a 15 minute walk from terminals 1/3.

4. South Side

A favourite for the photographers is a spot on the south side of the airport, alongside runway 23L/05R. Instead of turning into the Runway Visitor Park, continue along Wilmslow Road (A538) and through the tunnels underneath the runways. Immediately afterwards, turn left at the roundabout onto Altrincham Road. Park here, or in the adjacent hotel car park, then follow the footpath up through the woods towards the perimeter fence. Walk along the fence and find the raised hill right next to the runway. The sun is in a good position and the airliners are very close!

HOTELS

Radisson Blu Manchester Airport

Chicago Avenue, Manchester M90 3RA | +44 161 490 5000 | *www.radissonhotels.com*

The best hotel for spotting at Manchester. Located behind Terminal 2, rooms on high floors overlook the aprons and the runways in the distance. The restaurant also offers the same view. Some good opportunities for photographs with a long lens.

Public Transport: Manchester Airport has its own railway station with local and national links. Buses also link the airport to central Manchester, and can be used to reach the Runway Visitor Park (bus #288) and the south side viewing area (bus #88 from the RVP or terminals).

Newcastle

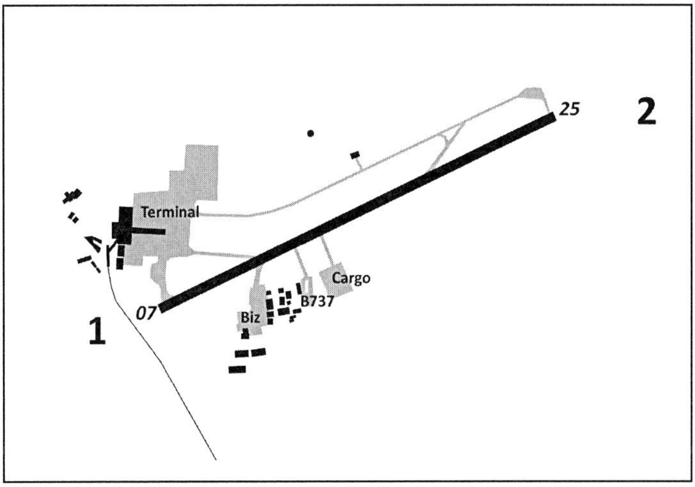

Runways: 07/25 7,641ft / 2,329m

Hub For: Jet2, Ryanair, TUI Airways

The busiest airport in North East England. Newcastle is a popular hub for holiday and low-cost flights, and a base for Jet2 and Ryanair. Other airliner traffic of note include a daily Emirates Boeing 777, Aer Lingus Regional, British Airways, Air France, Eurowings, Lufthansa, KLM, and Eastern Airways. It gets busier in the summer.

The passenger terminal is on the western side of the single runway. On the south side is a small cargo ramp which serves a few daily freighters, and a ramp for GA and executive aircraft. A local training college is also home to some airframes including a former Boeing 737-200 easily visible outside.

The best opportunities for spotting are at either end of the runway as no views exist landside in the terminal. At the southern end (runway 07), walk from the terminal along the B6918 until you reach the approach lights and metal gate, which leads to a small field where photographs can be taken (1).

At the opposite end (runway 25) the car park of a nature reserve is perfect for parking up to watch and photograph arrivals from the east (2). You can't see any aircraft on the ground from here, however, it is good for 25 arrivals. This is a short drive from the terminal, via Dinnington. Head for Coach Lane (postcode NE13 7BP).

Public Transport: Newcastle Airport is easy to reach on the Tyne & Wear Metro. The nature reserve spot is on bus route #44 from Regent Centre Metro stop.

Newquay

Runways

12/30 9,003ft / 2,744m

This fairly busy regional airport and former RAF base is situated in the south west corner of Britain. Flights are operated to UK and Ireland destinations, with the busiest period during the summer months when holidaymakers descend upon the region. There is also a popular link to the Scilly Isles.

The passenger terminal is on the north side of the runway, whilst the south side and former military section of the airfield has other aviation uses. One is the Draken Helicopter Academy, and a small FBO. Another is the Cornwall Aviation Heritage Centre, which has an interesting collection of (often active) old military aircraft and transport types. An aircraft maintenance company also operates on the south side, with a collection of airliners in storage and being parted out.

Spotting is possible in the West Car Park next to the passenger terminal. It is also possible to watch arrivals from a layby off the A3059 at the 30 end of the runway.

Public Transport: Bus #56 links central Newquay and Padstow with the airport passenger terminal.

The **Cornwall Aviation Heritage Centre** has a collection of historic aircraft, including a VC-10, BAC One-Eleven, Harrier, Hunters, Sea Devon and Varsity. Open Sunday to Friday from 10am, www.cornwallaviationhc.co.uk ** At the time of writing, notice had been given to the centre to vacate the site.

North Weald

Runways

02/20 6,171ft / 1,881m
13/31 2,703ft / 827m

An interesting general aviation airfield near Epping, between London and Chelmsford. North Weald Airfield was an important Battle of Britain station and many of the remaining buildings on the site are listed. Today it is a busy gateway for private aircraft flights, and is home to a number of vintage aircraft, including a Douglas C-54 Skymaster.

There are two cafes on site, the Wings Café and Rosey Lea Café, both good places to watch movements from. Access is by requesting permission from the gatehouse, which then involves driving around the taxiway. Beware of parked and taxiing aircraft! From here you can see most of the aircraft and any movements. Access is via Merlin Way (postcode CM16 6HR).

Public Transport: Buses #381, #419 and #420 all link Epping with North Weald Airfield

Norwich

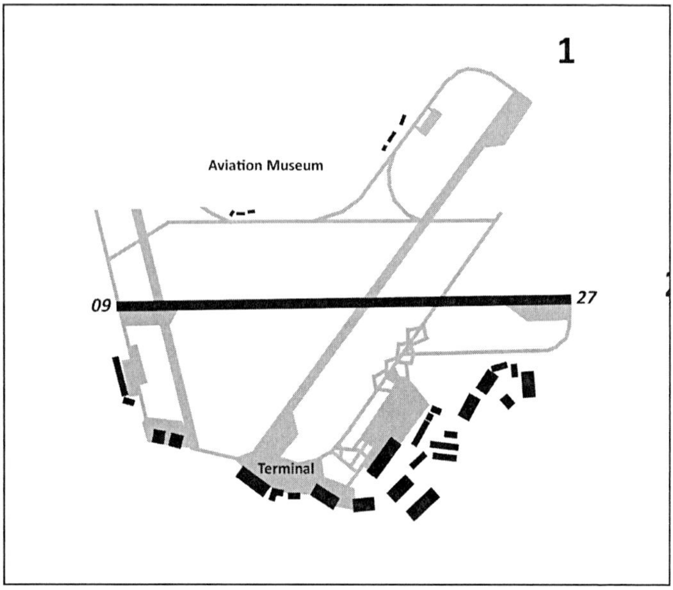

Runways

09/27 6,040ft / 1,841m

Norwich Airport handles around half a million passengers per year, mostly on scheduled flights KLM Cityhopper, Loganair and TUI Airways. In the summer months seasonal holiday flights operate, and offshore helicopter are plentiful, using their own facilities to the west of the terminal.

The airport is also home to a KLM Engineering base, so sees the Cityhopper fleet passing through regularly, and other airliners come in for painting and maintenance fairly frequently.

A selection of stored airliners is sometimes present on the north side of the airport, which in the past have included some interesting airframes. You can see these from the roundabout on the A1270 (1) on the north side of the airport (set SatNav to NR10 3GJ).

The car park and roads around the terminal are usually sufficient for seeing any aircraft parked on the aprons. There are also places to stop and watch arrivals on runway 27. Head to St. Faith's Road (postcode NR12 7BH) (2).

The **City of Norwich Aviation Museum** occupies a site on the northern boundary of the airport. It has a number of historic aircraft, including both a Fokker F-27 and Handley Page Herald of Air UK, which have significance to the airport. The museum is open daily in the summer and Wednesday, Saturday, Sunday and public holidays in the winter. See www.cnam.org.uk

Public Transport: Park and Ride buses link Norwich Airport with the city centre and train station. Local buses also pass nearby.

Old Warden

Runway

03/21 1,667ft / 508m

Old Warden Aerodrome is synonymous with the Shuttleworth Collection – one of the most important and historic collections of aircraft in the world.

The airfield is located 7 miles from Bedford and is easily accessed via the A1 motorway. Its main runway and taxiways are all of grass, with concrete parking aprons alongside the hangars to the north west.

The Shuttleworth Collection (see Museums section) puts on events every year, but you can visit the collection in normal opening hours.

Oxford

Runways

01/19 5,092ft / 1,552m

Oxford Airport occupies a fairly compact site 7 miles north of the city, next to the town of Kidlington. It no longer has passenger services, but is kept busy with executive aircraft movements, and as the home of CAE Oxford – one of the largest pilot training organisation of its kind in Europe. Its fleet of single- and twin-engine aircraft are on the go most days.

Langford Lane (postcode OX5 1RY) runs past the southern end of the main runway. There is a small layby to park in from where you can walk to the crash gate or back along the A44 to look across the parking aprons.

Driving into the airport, there are some views through fences and gates over the parking aprons and runway. You can park outside the Aviation Station Café. Security have been known to move people on from this area.

Public Transport: Bus #2C links central Oxford with the airport, via Kidlington.

Retford Gamston Airport

Runways

03/21 5,522ft / 1,683m
14/32 2,764ft / 842m

Although it has no passenger service, Retford Gamston is a busy general aviation site and sees occasional biz jet visitors. There are numerous flying clubs based here, and Diamond Executive Aircraft have a base. It is often possible to gain permission to walk airside along the line of hangars to log and photograph aircraft, and the café has a good reputation and decent views.

Southampton

Runways

02/20 5,653ft / 1,723m

Hub For: Eastern Airways, BA Cityflyer, Blue Islands

Southampton Airport is fairly busy as a regional airport, but has always been limited by its runway length. This is due to be rectified by an extension. Nevertheless, a number of summer services are also operated. Additional scheduled services are flown by Aer Lingus Regional, Aurigny, KLM Cityhopper and Loganair.

The passenger terminal and FBOs are on the western side of the runway.

There are two good places to watch aircraft at Southampton, but security are known to move people on from both. The first is the top floor of the multi-storey car park alongside the terminal and railway station.

The second viewpoint is reached by walking north along Mitchell Way towards the long stay parking. Past the hangars there is a raised area on your right which looks over the fence line to the runway and is great for photography.

Public Transport: Southampton Airport has its own railway station outside the terminal, linked to the South West Trains network.

Teesside International

Runways

05/23 7,516ft / 2,291m

Situated to the east of Darlington and the west of Stockton-on-Tees. KLM Cityhopper and Loganair provide daily scheduled flights, with various weekly Ryanair services and some summer charter routes. The airport is a popular base for GA flights and is a base for Draken with its Dassault Falcon 20 fleet.

Most aircraft can be seen by driving or walking the road behind the hangars. Photography is possible from the railway station bridge beyond the flying clubs, or the layby off the A67 near the end of runway 23.

On the south east side of the airfield is a fire training ground with a complete Trident 3B and wrecks of a Viscount, ATR42 and Shorts 330. It is visible from the railway station bridge.

A bar in the terminal offers views, with a viewing terrace planned for 2023.

Public Transport: Arriva bus #12 links the airport with Darlington.

Warton

Runway

07/25 7,946ft / 2,422m

Warton Aerodrome is in Lancashire, not far from Blackpool and Preston. It is a private airport, operated by BAe Systems where traditionally both military and civilian aircraft were manufactured. The Eurofighter Typhoon is assembled here.

Today the airport mostly sees movements of corporate transport aircraft ferrying workers between sites, often operated by Embraer 145 aircraft.

The best views are from Stoney Land in the south-east corner of the airfield, where you can see through the fence to the runway 25 threshold. You'll need a ladder for decent photographs. Head for postcode PR4 1RA.

Wellesbourne Mountford

Runways

05/23 1,926ft / 587m
18/36 3,009ft / 917m
18G/36G 1,660ft / 506m

This is a busy little general aviation airfield in Warwickshire, close to Stratford-upon-Avon. Attempts to close it and redevelop the site in recent years have so far been thwarted, but it may not be the end of the story.

The airfield has an active Avro Vulcan bomber on site which sometimes performs taxi runs. The best views of this, and all other resident aircraft, can be had by driving along Loxley Lane (postcode CV35 9EU) which passes the flying clubs, café, small museum and helicopter centre. You can pull into the car park at each to log what is visible.

London

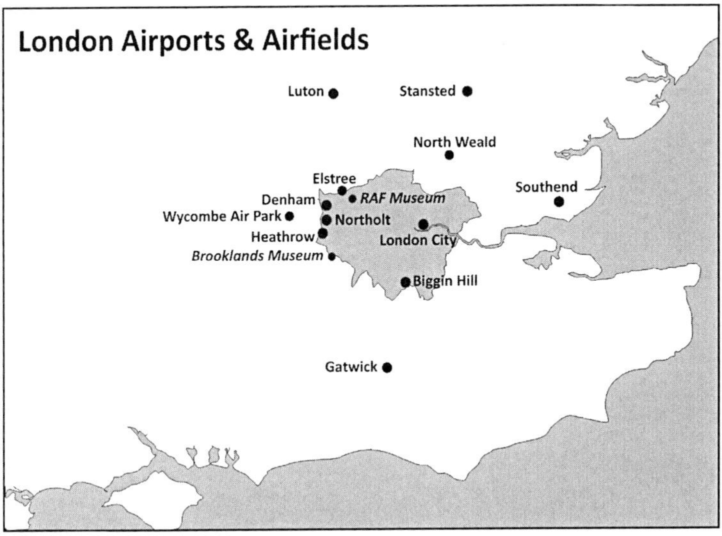

Deserving its own section, London is the busiest airspace in the United Kingdom with the highest density of major airports and significant airfields. The map here shows the location of those listed below in relation to the city itself; transport links to all are fairly good, and the aircraft traffic is also noteworthy to the enthusiast.

Additional aviation attractions in and around London include the excellent Brooklands Museum at Weybridge, and the Royal Air Force Museum at Hendon, both also labelled on the area map.

Denham

Runways

06/24 2,543ft / 775m
12/30 1,791ft / 546m

Denham Aerodrome is just inside the M25 London orbital motorway, close to where the M40 intersects. It is 13 miles north of Heathrow Airport. Denham has a concrete and a grass runway and is a busy general aviation airfield for London. Light aircraft and helicopters make up most movements, with the occasional biz jet or prop.

The airfield is split into north and south sections, with Tilehouse Lane linking them. The north side is home to The Pilot Centre flying school, who are known to allow airside access for logging and photography if you have a hi-viz vest. There's also a café with views here. The south side has a parking area alongside the fence and a restaurant with views.

Public Transport: Denham Aerodrome is a 30 minute walk from the nearby Denham Station and bus routes.

Elstree

Runway

08/26 2,136ft / 651m

Elstree is a busy little general aviation airfield to the north of London, just inside the M25 motorway and alongside the M1 as it leaves London.

The airfield has a single concrete runway, with parking aprons, hangars, café etc to the south. Views are possible from the café, or ask at the control tower to walk out and log aircraft.

London Biggin Hill

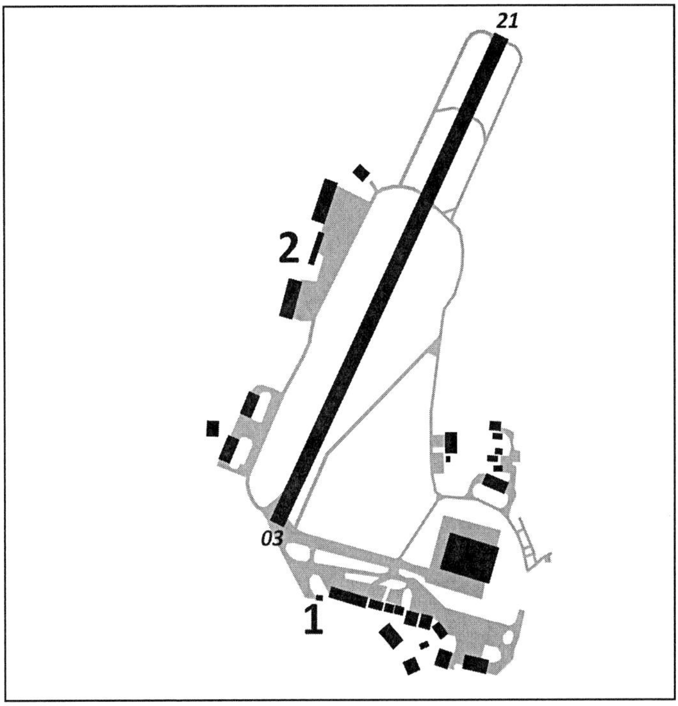

Runways

03/21 5,971ft / 1,820m

Biggin Hill was a wartime base famous for its involvement in the Battle of Britain. Located near Bromley, some 14 miles south of central London, today it acts as a gateway for biz jets and general aviation traffic. It currently has a ban on scheduled services, preventing it from entering the competitive market for flights to London, but nevertheless sees aircraft up to the Airbus A320 and Boeing 737 size, albeit as executive transports.

Various FBOs are located at Biggin Hill, using hangars and aprons along the western and southern boundaries of the airport. The passenger terminal is on the western side, off the A233. On the eastern side are hangars and aprons generally used by light aircraft and companies involved in the restoration of heritage aircraft.

The Lookout Coffee Shop (1) offers a place to enjoy a drink whilst watching aircraft movements at Biggin Hill. You can find it on Maitland View (postcode TN16 3BN). From here you will see most movements, particularly on the main runway and nearby parking areas.

A security barrier prevents you from exploring any deeper within the airport complex without a good reason. Alternatively you can try outside the passenger terminal (2) for a quick view of parked aircraft.

Public Transport: Bus #320 links Biggin Hill with Bromley railway stations, which have links to London.

London City

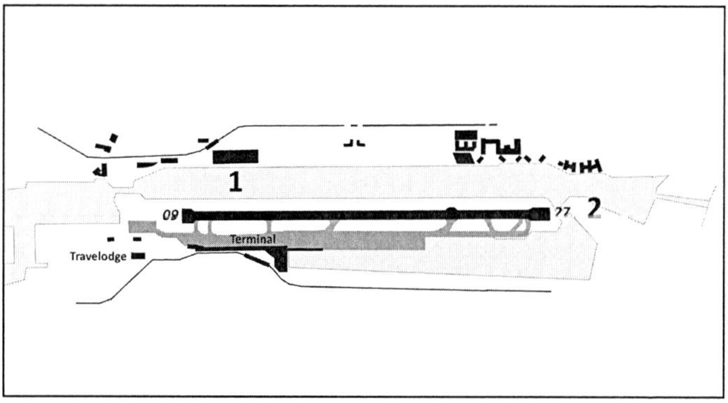

Runways

09/27 4,948ft / 1,508m

Hub For: BA CityFlyer

At the heart of London's Docklands, built on a former dock, is the compact London City Airport. Despite its size and short runway, this is a busy hub and even had transatlantic services until recently. The airport is a hub for BA Cityflyer (a British Airways partner), and is served from Europe's capitals and business destinations by ITA Airways, KLM Cityhopper, Loganair, LOT Polish Airlines, Lufthansa Regional, Luxair, and Swiss International Air Lines.

Given the restrictions on runway length and the steep approach angle owing to nearby skyscrapers, the airport's movements are mostly limited to regional jets and turboprops.

The airport has been undergoing expansion of its terminal, parking apron and taxiways which will likely see more flights introduced. At present the airport is busiest early in the morning and late in the evening. It closes for flights over a 24-hour period between 1pm Saturday and 12.30pm Sunday.

The terminal building and apron are on the south side of the runway, with a dedicated ramp for biz jets to the east.

SPOTTING LOCATIONS

1. Opposite Dock

On the dock opposite the airport to the north, there is a walkway running the length of the runway with great views. From here you will see every movement and be able to read off aircraft parked at the terminal. Some parked biz jets are a little awkward to read from here, however. Photography is also great from this spot with a 300mm lens. To get here head for Royal Albert Way, off the A1020, or walk 20 minutes across the Connaught Bridge from the terminal; it is also the Royal Albert stop on the DLR. The University of East London is here, and there is ample parking space.

2. Runway 27 Lookout

When heading north on the A112, there is a small industrial estate just across from the runway 27 end. Take a right turn into the estate onto Gallions Road, and then turn left and follow the road under the A112 to a small car park. This spot is very close to the end of Runway 27, so you get a good view down the runway which is perfect for photography (postcode E16 2GG). It's a 25 minute walk from the terminal, or the Cyprus DLR station.

HOTELS

DoubleTree by Hilton London ExCeL

2 Festoon Way, Royal Docks, London E16 1RH | +44 20 7540 4820 | *www.hilton.com*

On the north side of the dock near the ExCeL Centre. The best views are from the top (sixth) floor even-numbered rooms. You'll be able to see the threshold of runway 09, with approaching aircraft passing by the room. Departures from 27 are also visible.

London City Airport Travelodge

Hartmann Road, Silvertown, London E16 2BZ | 0871 984 6290 | *www.travelodge.co.uk*

A short distance from the terminal and on the main access road. Higher rooms facing north have a view over the end of runway 09 and the biz jet ramp. The best ones are 503 to 509. You shouldn't miss too many movements from here. Please note that these are family rooms, so you'll need to request one when booking.

Public Transport: The airport is served by buses and the Docklands Light Railway which links to the London Underground and mainline train stations.

London Gatwick

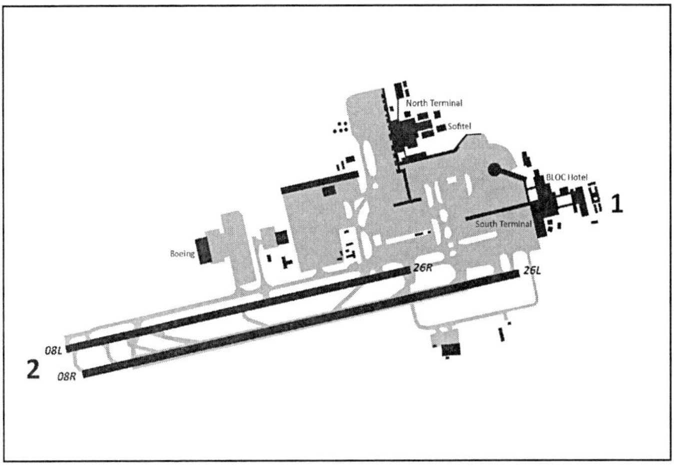

Runways

08L/26R 8,415ft / 2,565m
08R/26L 10,879ft / 3,316m

Hub For: British Airways, easyJet, Norse Atlantic Airways, Norwegian Air Shuttle, Ryanair, TUI Airways, Wizz Air

Gatwick, London's second largest airport, is situated to the south of the city next to the town of Crawley. Whilst it is the world's busiest single-runway airport operation, you'll note that a second runway does exist, albeit acting as a taxiway except when the main strip is out of operation. Thus Gatwick can appear a crowded and busy place during peak hours.

Much has changed at Gatwick since the Covid-19 pandemic, with the loss of much of British Airways' network (slowly being rebuilt), Virgin Atlantic's base, and Norwegian's long-haul network (now resurrected under the Norse Atlantic brand).

Nevertheless, easyJet, TUI and Wizz Air maintain a strong network here, and many other leisure, low-cost and long-haul airlines are also present. Of the long-haul carriers, Emirates, JetBlue, Qatar Airways and WestJet are probably the most interesting.

Gatwick has two terminals – North and South. There is a fairly even split of carriers across the two terminals, with easyJet and TUI using both, and British Airways and Wizz Air in the South Terminal. There is an interesting passenger bridge over the taxiway at the North Terminal, with airliners passing below.

Maintenance areas exist on the south side of the airport (with a resident MD-80 airframe), and a new Boeing facility to the north west of the runways.

Sadly Gatwick lost its excellent viewing terrace on the South Terminal in favour of expansion in the early 2000s. This left the airport with no official viewing facilities, and it has since proved a very frustrating airport for the enthusiast. For those with a flight ticket, there are good places to spot particularly in the South Terminal. The BLOC Hotel is also good.

SPOTTING LOCATIONS

1. Blue Car Park

The top level of the multi-storey Blue car park outside the South Terminal is a nice spot for logging aircraft on short finals to Runway 26L. Facing into the sun is not ideal, however. Signs at this location indicate that spotters are not welcome to loiter. A similar view exists from the platforms of the adjacent airport railway station.

2. Runway 08R

Following Charlwood Road and Lowfield Heath Road (postcode RH11 0QB) around the end of Runway 08R leads to a crash gate which is close to aircraft lining up on the runway. However, it is no longer possible to photograph through the gate. An alternative is the area around the Purple Parking car park to the south. Parking is difficult, so it may be best to walk, or catch the Purple Parking bus. The area is good for photographing runway 08R arrivals, but not those on the ground.

HOTELS

BLOC Hotel

South Terminal Gatwick Airport, RH6 0NP | +44 20 3051 0101 | *www.blochotels.com*

Situated atop the South Terminal in the former administration building, the BLOC Hotel is a great place to spot if you have a room facing the airport. It has similar views to the old viewing terrace. Depending on the room you will usually have a view of both terminals and part of the runway.

Sofitel London Gatwick Airport

North Terminal, Gatwick Airport RH6 0PH | +44 1293 567070 | *www.sofitel.com*

A smart hotel situated at the North Terminal, and linked via monorail from the South Terminal. Rooms on the higher floors facing the airport have unrivalled views of aircraft movements to both terminals.

Public Transport: Gatwick Airport has a rail station outside the South Terminal with fast train services to central London and Brighton. Catch bus #26 from the South Terminal (also stops near the North Terminal) to Charlwood, which is next to the Gatwick Aviation Museum, and a ten minute walk from spotting location 2.

The **Gatwick Aviation Museum** at Charlwood, close to the end of runway 08L/R at Gatwick houses a nice little collection of military jets and historic aircraft. It is open Friday, Saturday, Sunday 10am to 4.30pm, www.gatwick-aviation-museum.co.uk

London Heathrow

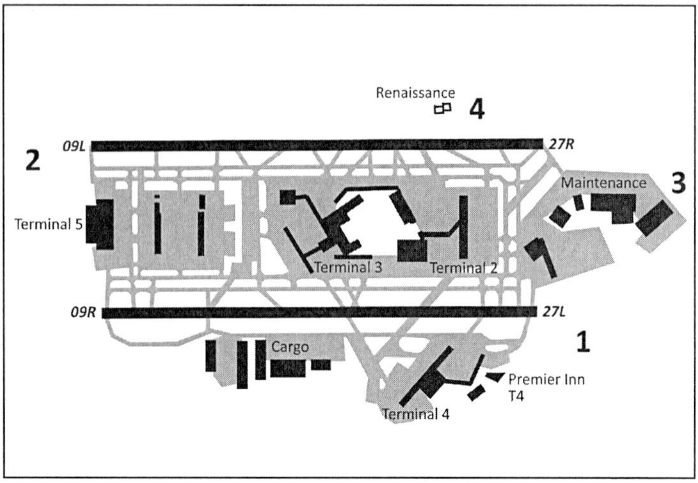

Runways

09L/27R 12,802ft / 3,902m
09R/27L 12,008ft / 3,660m

Hub for: British Airways, Virgin Atlantic

Heathrow is the country's busiest airport and at times is very overcrowded. A decision over a new runway at the airport has been held up many times, leaving Heathrow to struggle with capacity and ambitions for growth. Nevertheless, the variety of airlines operating here can be mouth-watering to the enthusiast, coming from all corners of the globe. It is the operating and maintenance base for both British Airways and Virgin Atlantic, and features as a destination for many world carriers.

Heathrow has two parallel runways, 09L/27R and 09R/27L. Patterns of runway assignment between landing and departing switch at 3pm every day when in the 27 direction, to relieve local neighbours. When using the 09 direction, arrivals are always on 09L and departures on 09R.

The airport has four terminals in use. Terminals 2 and 3 are in the central area, with 4 to the south east and 5 to the west. Terminal

5 is occupied primarily by British Airways and its partners. Cargo aircraft park to the south, and maintenance areas are to the east.

Heathrow sees large numbers of widebody jets amongst the airlines flying in, with relatively few smaller regional jets and props. Whilst there are dedicated cargo movements, most freight is carried in the belly of passenger aircraft.

SPOTTING LOCATIONS

1. Myrtle Avenue

This is one of the most popular spots at Heathrow, but is only useful when aircraft are landing on runway 27L. The spot gets its name from a small residential street close to Hatton Cross, with a grass area at the end. Spotters congregate on this area to log and photograph aircraft as they pass low overhead. The postcode is TW14 9QU. There is very limited parking, so it's best to walk from Hatton Cross Tube Station.

2. Runway 09L Approach

When aircraft are arriving on runway 09L, a good spot is outside Terminal 5. At ground level, walk north out of the bus bays along Wallis Road (there's a footpath). It will pass a small security point, and on the opposite side of the road is a grass embankment, with narrow paths where pipes run through. Climb one of these to the area of grass under the trees at the top. Here you'll have good opportunities for logging and photographing arrivals on this runway, but see little else at the airport.

3. Eastern Perimeter Road

When arrivals are on 27R, a walk along the Eastern Perimeter Road can help you find a good spot to watch arrivals on 27R. You can start at either Hatton Cross Underground Station, or the Northern Perimeter Road. Security patrols regularly monitor this area and ask spotters to stay a few metres from the fence. You can take good photographs here of arrivals, but can't see any other parts of the airport.

4. Academy/Renaissance Hotel

Even if you're not staying in the Renaissance Hotel (see below), the car park next to it was formerly the home of the Heathrow Visitors Centre. Today it is the Heathrow Academy and has an enthusiast's shop. From here you can get good views of all movements on the northern runway (09L/27R) from a small grandstand. It is not the best place for photographs. This spot is located off the Northern Perimeter Road at postcode TW6 2RE.

HOTELS

Renaissance London Heathrow

140 Bath Road, Hounslow TW6 2AQ |
+44 20 88 97 63 63 | *www.marriott.com*

This is one of the best spotting hotels in the world. Request a room overlooking the airport (you sometimes have to pay more). All movements on the northern runway can be read off and photographed easily, but lower floors have lamp posts and the fence in the way. Movements around the terminals are easy to spot. Those using flight tracking websites can continue to spot throughout the night. Although this hotel is not the cheapest at Heathrow, the quality of spotting makes up for it and it offers special spotter packages through its website.

Premier Inn Terminal 4

Sheffield Rod, Heathrow TW6 3AF | +44 (0) 333 234 6600 |
www.premierinn.com

A relatively new hotel, situated near the entrance to Terminal 4 on the south eastern corner of the airport, close to Hatton Cross and the Myrtle Avenue spotting location. Rooms on the 6th and 7th floor facing the runway have excellent views of aircraft using runway 27L, with Terminal 2 and runway 27R in the distance. Photography possible through two sets of double glazing. You won't miss much here.

Hilton Garden Inn London Heathrow Terminals 2 and 3

Heathrow Airport, Terminal 2, Inner Ring E, Longford, Hounslow TW6 1AH | +44 20 3972 9700 | *www.hilton.com*

A brand new hotel, and the only one in the central terminal complex. It is tall, with a top floor Runway Bar offering 360 degree views around the airport. Many rooms offer views of the action, but at a distance or with buildings blocking part of the view. Ask for a higher level room facing Terminal 3 and you'll get a good view of many of the gates, and runway 09R/27L beyond.

Thistle London Heathrow

Bath Rd, Longford, London UB7 0EQ | +44 20 7523 5056 | *www.thistle.com*

A good option if aircraft are landing on runway 09L from the west. This hotel is opposite the threshold of this runway, close to Terminal 5. Few rooms have views, but you might get lucky if you ask. There is a terrace you can use which has good views.

Public Transport: Naturally Heathrow is well served by public transport links from London and surrounding areas, plus the national bus network. You can reach all terminals and nearby Hatton Cross (good for spotting location 1) on the London Underground Piccadilly line. The airport is also linked to Paddington Station by the Heathrow Express train.

Buses run around the perimeter roads and between terminals. Pay the driver. You can reach spotting locations 3 and 4 using bus #285 from the central terminal area, or bus #423 from Terminal 5.

London Luton

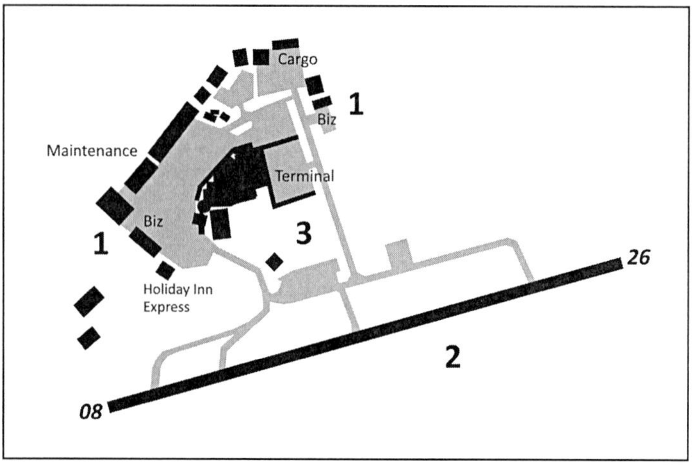

Runways

08/26 7,087ft / 2,162m

Hub For: easyJet, Ryanair, TUI Airways, Wizz Air

Luton is London's fourth largest airport, and is situated some 30 miles north off the main M1 motorway. The airport is a busy gateway for low-cost carriers and holiday charter airlines. It is the home base of easyJet, TUI Airways and Wizz Air UK. Ryanair also has a heavy presence. A number of cargo airliners pass through most days.

Luton is perhaps best known amongst enthusiasts for the large variety of business jets which pass through on a regular basis. The airport has many ramps and hangars dedicated to this traffic, which can be a little difficult to navigate around to see everything. However, the quality is often worth it, with aircraft from far off places commonplace.

Luton has a cramped and confusing central area which includes the terminals, car parks, roads, hotels and administration buildings. You will see signs around this area discouraging spotters from stopping. If you have some time to spend here, it's best to spend it at the viewing area alongside the runway. Otherwise a quick stop to log through the fence should be fine.

SPOTTING LOCATIONS

1. Central Area

Driving around the access roads to the hangars, cargo centre and other parts of the central complex will offer glimpses of many business jets and other aircraft parked around the various aprons. Signs discourage spotters from parking in these areas, so make only quick stops to log what you can see. You can park at the Holiday Inn for the nearer ramps, and you can use the Long Stay Term car park for free up to an hour (follow the road as far as you can) to log the biz jets parked alongside it.

2. Crash Gate

The postcode for the crash gate is LU2 8PE. From the central terminal area it is a complicated route along country lanes via the village of Beechwood Green. At the end of the lane leading to the crash gate, there is a small area for parking. Keep the crash gate clear at all times! Views of aircraft using the runway are good from here, and photography is possible.

3. Multi-Storey Car Park

The relatively new multi-storey Car Park outside the passenger terminal has good views from the top level over the taxiways, runway and some of the executive parking aprons. Parking charges apply.

HOTEL

Holiday Inn Express Luton Airport

2 Percival Way, Luton LU2 9GP | +44 871 902 1622 | *www.ihg.com*

Rooms facing the airport are all great for logging aircraft on the runway and some taxiways, and also have plenty of opportunities for good photographs. If you're facing the other direction you have a view of part of the passenger and corporate parking areas.

Public Transport: Buses #99, #757, #A and National Express coaches all link central Luton with the airport, which is a short distance. The free bus to the Long Term Car Park is a good option for reaching the spotting locations there.

London Southend

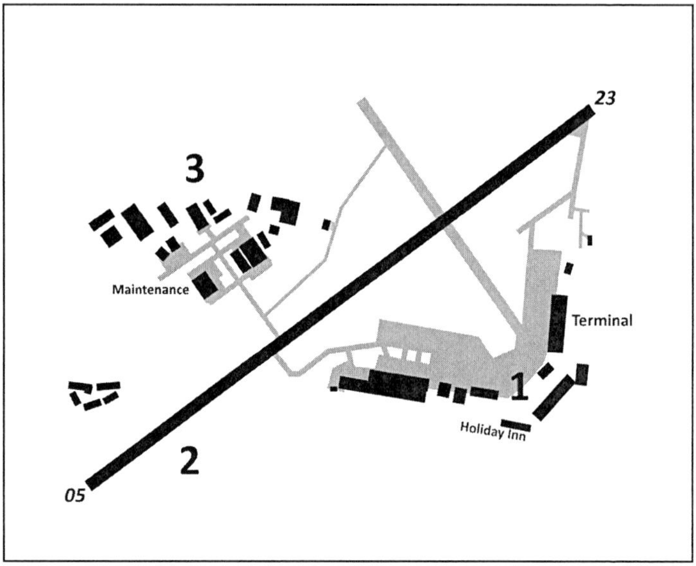

Runways

05/23 6,089ft / 1,856m

Rebranded like many periphery airports around the capital to take advantage of passenger traffic from the significance of serving London. Southend is in fact over 40 miles to the east in Essex, but has up to eight trains per hour linking it to Liverpool Street Station in around 50 minutes.

Southend Airport was long an enthusiast favourite because of the interesting mix of aircraft usually present, and the memories of early independent carriers which would ply their trade here. However, these days are largely gone, and in recent years under new management, Southend has focussed more on passenger services having attracted the likes of easyJet and Flybe to boost the number of passengers passing through its new terminal building and runway extension.

easyJet's base closed in 2020, but it still operates seasonal service along with some other carriers. Cargo movements can also be seen.

The vast majority of Southend's traffic today is general aviation. Any aircraft undergoing maintenance are usually parked around the hangars on the north side of the airport.

SPOTTING LOCATIONS

1. Car Parks

The southern part of the passenger terminal car parks have views across parts of the ramp.

2. Runway Viewing Area

When the runway was extended the main road passing behind it was truncated, leaving small sections on both the south and north sides which have views onto the runway and small areas to park up. No entry signs discourage venturing too close, and parking is difficult.

3. Aviation Way

This road leads behind the various hangars and warehouses on the northern side of the airport. It is where most of the stored aircraft and those receiving maintenance can be found, and a drive along here uncovers various vantage points. Postcode SS2 6UN.

HOTELS

Holiday Inn Southend Airport

77 Eastwoodbury Cres, Southend-on-Sea SS2 6XG | +44 1702 543001 | *www.holidayinn.com*

This hotel's location is perfectly suited for in-room spotting. All rooms on the north side of the building (including Standard, Executive, and King Superior rooms) face towards the runway and terminal ramp. The rooftop restaurant and bar also has panoramic views.

Public Transport: Train services link the passenger terminal with London Liverpool Street Station. Bus #9 from the passenger terminal routes past the end of the runway and spotting location 2.

London Stansted

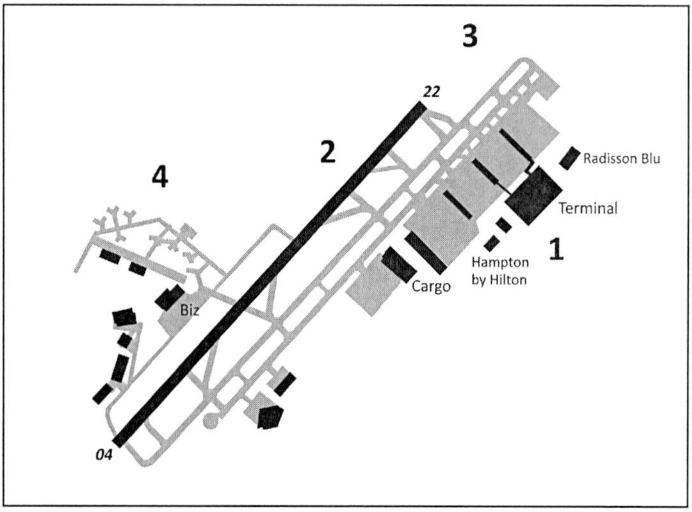

Runways

04/22 10,003ft / 3,049m

Hub For: easyJet, FedEx Express, Jet2, Ryanair, Titan Airways, TUI Airways

Stansted is a fairly busy airport 30 miles north east of London which is popular with low-cost carriers and cargo airlines.

The airport had started life as a bomber base and maintenance depot in the Second World War. The potential for passenger operations were realised when the British Airports Authority (BAA) took over in 1966.

When low-cost airlines Ryanair, Go and Buzz wanted cheaper access to London, Stansted was chosen and it quickly grew to handle the demand, having opened a brand-new terminal in the early 1990s. Today many low-cost and leisure airlines have a presence at

Stansted, however it is a much quieter place than pre-Covid. The only remaining long-haul carrier is Emirates.

Stansted has a single runway. On the north side extensive aprons and hangars for executive aircraft movements can be found.

The passenger terminal has three concourses of gates which are not visible from the rest of the terminal. Further to the west are two areas of cargo aprons and warehouses where the many daily flights park up. These include carriers such as Asiana, Cargolux, China Southern, FedEx Express, Turkish Airlines Cargo and UPS Airlines. Ryanair also has a maintenance base here.

SPOTTING LOCATIONS

1. Old Terminal Area

On the opposite side of the airport close to the runway 04 end is the old terminal area, now used for maintenance, executive jets, and aircraft storage. On any day there are likely to be some exotic business jets present. When exiting the M11 towards the airport, follow signs for Long Stay car parks and Business Park, following Round Coppice Road. Turn right onto First Avenue (postcode CM24 1RY) into the business park and look for places to quickly log what you can see without blocking any roads. Beware that security do not like spotters being here or looking through the fences, so caution and haste is advised.

2. Belmer Road Viewing Area

Drive back and turn right to continue along Bury Lodge Lane. In the tiny village of Burton End turn left onto Belmer Road (use postcode CM24 8UL). After half a mile find a place to park by the side of the road, and then walk through the paths in the trees to find an official viewing area alongside the fence, opposite the runway. You have distant views of the passenger and cargo terminals, but all runway movements are easy to see, with photography through the fence.

3. Green Street

At the north eastern end of the airport, this spot is good for arrivals on runway 22. From the main roundabout at the entrance to the passenger terminal, take the exit for Molehill Green/Broxted/Elsenham along Hall Road. After passing the runway end and a small area of houses and industrial units, look for a left turning onto Green Street. Follow it until it becomes a track and reaches a crash gate.

4. Ash Pub Track

Before reaching the Belmer Road spot (location 2), you'll pass a pub called The Ash. It has limited parking available – intended for customers. Alongside the pub is a track used as a public footpath by dog walkers etc. Following it leads to some vantage points looking through the fence around the northern area.

HOTELS

Hampton by Hilton London Stansted

Bassingbourn Rd, Stansted CM24 1QW | +44 1279 701101 | *www.hilton.com*

A recent addition to Stansted, to the south of the main terminal. Rooms in the even-numbered 634-658 and 734-758 range have a panoramic view over most of the terminal and part of the cargo area. You will also see all runway movements. Photography is easy, through glass, until sun moves over on the afternoon.

Radisson Blu London Stansted Airport

Waltham Close, Stansted Airport, Essex CM24 1PP | +44 1279 661012 | *www.radissonblue.co.uk*

A very smart, modern hotel at Stansted with prices at the higher end. Some higher rooms facing the airport have views over the Ryanair pier and runway 22 threshold. The hotel is only a few metres from the Terminal.

Public Transport: The passenger terminal has a railway station linking to London, and buses arrive from many places. It's not possible to get to either of the spotting locations via public transport.

Wycombe Air Park

Runways

06/24 2,411ft / 735m
06G/24G 2,001ft / 610m
35 2,280ft / 695m

Wycombe Air Park is an airfield at Booker in Buckinghamshire, to the west of London and north of Reading. It is often known as Booker Airfield.

The airfield was established in World War II, but today is a busy general aviation field with hundreds of daily movements – mostly through based training aircraft.

Wycombe is located just off the M40 motorway. Drive to the main car park at the airfield and you'll be able to see most aircraft parked outside.

England's Military Airfields

RAF Barkston Heath

Runways

06/24 6,129ft / 1,868m
10/28 4,324ft / 1,318m
18/36 4,199ft / 1,280m

Base for: No 703 Naval Air Squadron

Located near Grantham in Lincolnshire, Barkston Heath is a training base for the Naval Element of No. 3 Flying Training School RAF (No. 3 FTS), providing elementary training for the Royal Navy using Grob Prefect T1 aircraft.

The main entrance, off High Dyke (postcode NG32 3FH) has space to quickly pull in and log aircraft parked on the main apron.

MOD Boscombe Down

Runways

05/23 10,515ft / 3,205m
17/35 6,864ft / 2,092m

Base for: Aircraft Test & Evaluation Centre, Empire Test Pilots School, Rotary Wing Test & Evaluation Squadron, 744 Naval Air Squadron, Bristol University Air Squadron, Southampton University Air Squadron, Joint Aircraft Recovery and Transportation Squadron, QinetiQ

MOD Boscombe Down, operated by QinetiQ, is a former RAF airfield at Amesbury in Wiltshire. It has two active runways and a large operational area along the northern part of the site.

Various aircraft operate here at different times, often in evaluation with one of the based test centres. QinetiQ also has a selection of larger aircraft based here. There is also a Lightning aircraft as a gate guardian.

The A345 runs down the western side of the airfield. Head south, then turn left onto Winterslow Road, which has a small viewing area close to the end of runway 05 where there's room to park. The fence is high, so you'll need to be creative for photographs.

An area also exists at the 23 end. From the A303 heading east from Amesbury, turn right down a narrow country lane (you'll need to cross the other carriageway). This track leads down to an area looking down the runway where you can see all movements and some of the parking aprons.

RAF Brize Norton

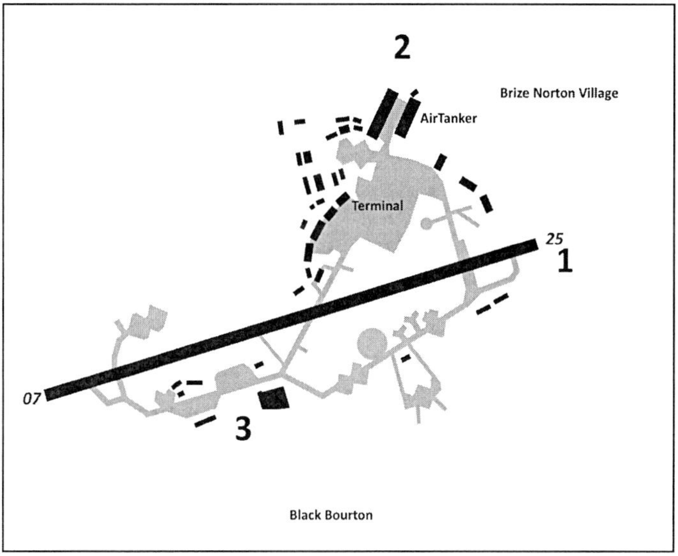

Runway

07/25 10,007ft / 3,050m

Hub For: Royal Air Force

Brize Norton in Oxfordshire is the Royal Air Force's principal air transport and air logistics base. It is home to the AirTanker fleet of Airbus A330 Voyagers, plus military parachuting and the air link to the Falkland Islands and Ascension. In addition the RAF's Lockheed C-130 Hercules, Airbus A400M and Boeing C-17 Globemaster III fleets operate from here. Most weeks a variety of support aircraft, from biz jets to airliners and cargo aircraft, pass through.

The airfield has one runway running east-west with aprons and technical areas on both sides. The majority of the base, including the passenger terminal and AirTanker hangars are on the north side. Entering the base without a reason is generally not permitted, so any spotting needs to be done from outside the airfield.

SPOTTING LOCATIONS

1. Runway 25

One of the best locations for photography is the end of runway 25 where Station Road passes very close. You can't park here, so need to find a spot in Brize Norton village which is a 10 minute walk.

2. Carterton Road

You can walk along Carterton Road running west out of Brize Norton village, grabbing fleeting views of any larger aircraft parked up among the hangars. Photography not possible.

3. Southern Perimeter

Head for Black Bourton village, and then follow Burford Road (OX18 2PF) north until you reach the fence. Park here, being careful not to block the crash gate, and then walk west along the track which starts opposite the small distillery. It turns into the Black Bourton Footpath and eventually rounds the runway 08 end and emerges in Carterton village. Along the way you can stop off to see any aircraft parked on the southern dispersals and ramps.

Public Transport: Bus #S1 runs from nearby Oxford and Witney to Brize Norton village.

RAF Coningsby

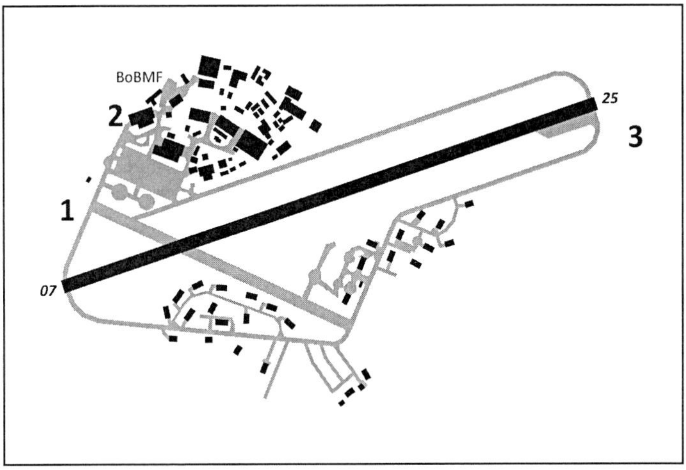

Runway

07/25 9,003ft / 2,744m

Base for: No 3 Squadron, No 11 Squadron, No 12 Squadron, No 29 Squadron, No 41 Squadron, Battle of Britain Memorial Flight

Coningsby, in Lincolnshire, is one of the most popular military airfields in Britain thanks to the presence of the Battle of Britain Memorial Flight (BoBMF) – a squadron of preserved and airworthy historic aircraft, including Spitfires, Hurricanes and one of only two airworthy Avro Lancasters in the world. These aircraft regularly take part in displays and fly-pasts around the country, and can often be seen in the air around this airfield.

Coningsby today is also home to Eurofighter squadrons, including a training unit, and is a Quick Reaction Alert (QRA) base.

Thankfully spotting at Coningsby is easy.

SPOTTING LOCATIONS

1. Spotters Car Park

Along Dogdyke Lane there is a car park situated at the end of runway 07 (postcode LN4 4TE). From here you can sit and watch or photograph movements on the runway (however 25 departures are a little high by this point).

2. BoBMF Layby

A little further up Dogdyke Lane is a layby next to the fence which looks out onto the Battle of Britain Memorial Flight hangar and apron. It's great for seeing these aircraft coming and going, and you can also see Typhoons parked up beyond.

3. Runway 25

The B1192 runs past the end of runway 25. To get there you can continue past spot 1, then turn left onto New York Road and left onto the B1192/Langrick Road, heading north. Just before the traffic lights at the end of the runway is some rough ground on the left, in a farmer's field. You can pull onto here and park (with a donations box for local charity). If aircraft are landing on 25 this is a great spot for photography.

Battle of Britain Memorial Flight Visitors Centre

Dogdyke Road, Coningsby, LN4 4SY | 01522 782040 | *bbmf@lincolnshire.gov.uk*

You can visit the hangar where the BoBMF aircraft are maintained and kept whilst not away on display. The centre is open Mon-Fri 10am-5pm, excluding Bank Holidays. Tours begin at 10.30am and the last one is at 3.30pm. Adults £9.80, Concessions £8.30, Children (5-16) £5.70, Family £26.80, Under 5s free.

RAF Cranwell

Runways

09/27 6,831ft / 2,082m
01/19 4,797ft / 1,462m

Base for: Central Flying School, No 3 Flying Training School, No 45 Squadron, No 57 Squadron, No 6 Flying Training School, No 7 Air Experience Flight

Home of the Royal Air Force's Central Flying School and Royal Air Force College, which are responsible for training and turning out new officers and aircrew, Cranwell in Lincolnshire is a busy air base. As such, most aircraft based here are turboprop and jet training aircraft, such as the Grob Prefect, Beech King Air and Embraer Phenom T1.

The main aircraft apron is in the north west corner of the airfield. The best general location for spotting is along the A17, which runs down the western side of the airfield. Adjacent to the end of runway 09 is a small parking area, where spotters usually congregate. You can see the main parking apron from here, and good photographs can be had of aircraft approaching 09. Further parking is available a little further south.

Rauceby Lane runs along the eastern side of the airfield, with good views of runway 27, but parking opportunities are very limited.

RNAS Culdrose

Runways

06/24 3,428ft / 1,045m
11/29 6,001ft / 1,829m
18/36 3,461ft / 1,055m

Base for: Fleet Air Arm

Culdrose is one of Europe's busiest helicopter bases. It is located at Helston on the Lizard Peninsula of Cornwall, in south west England.

Culdrose is also known as HMS Seahawk.

Its based operators include active Naval Air Squadrons, as well as engineering, maintenance and training operators, with the AgustaWestland Merlin helicopter the most common sight. You'll also see Navy Beechcraft King Air Avenger aircraft based here.

A public viewing area exists at the eastern end of the airfield, with parking and good views of movements, albeit a little distant from the western operational areas of the airfield. Follow the A3083 which passes the southern side of the airfield between Helston and Lizard, and turn onto the B3293. After a short distance turn left, signposted for the viewing area.

RAF Fairford

Runway

09/27 9,993ft / 3,046m

Base for: 99th Expeditionary Reconnaissance Squadron

Fairford in Gloucestershire is most famous the world over as the venue for the Royal International Air Tattoo (RIAT) held each summer, which draws hundreds of thousands of enthusiasts from around the world to witness an impressive array of air force (and even some civilian) aircraft both in the air and as part of the static display.

For most, this is the only time to visit the airfield, which can be quiet at other times.

However, throughout the year Fairford often has some interesting visitors, including hosted units from the United States Air Force.

The western side of the airfield has places to spot aircraft on approach to runway 09. It's best to park on the edge of the village of Marston Meysey and walk north. Use postcode SN6 6LQ.

You can also walk along the southern perimeter fence near the village of Dunfield (postcode GL7 4HE) where you can see aircraft on the southern parking aprons close up. But don't loiter.

RAF Lakenheath

Runway

06/24 8,999ft / 2,743m

Base for: 48th Fighter Wing

RAF Lakenheath is a host base for the United States Air Force (USAF) in Suffolk. It hosts the 48th Fighter Wing and its individual squadrons, which at present see based F-15 and F-35A Lightning aircraft based. The airfield is only a short distance from RAF Mildenhall (see below), another USAF base in south east England.

Only one runway is active today on this former wartime airfield. All aircraft operations centre on the area to the south of the runway.

An official viewing area is situated at the northern end of the runway, close to the 24 threshold. It is reached off the A1065. Just to the north of the airfield, turn left onto Wangford Road, then left again to the signposted viewing area (postcode IP27 0SJ).

Parking is provided, and you have a good view of movements. However, photography is usually into the sun, and a tall fence surrounds the viewing area. You may choose to walk back along the A1065 for less cluttered shots, with the sun behind you.

RAF Leeming

Runway

16/34 7,516ft / 2,291m

Base for: No 135 Expeditionary Air Wing

Leeming is a Royal Air Force base in North Yorkshire, close to Northallerton and just off the A1(M) at Leeming Bar. It is presently home to the joint Qatar Emiri-Royal Air Force Hawk Training Squadron, with based Hawks of the Qatar Emiri Air Force.

The airfield also often plays host to visiting squadrons during exercises.

A viewing area exists to the north-west of the airfield on Roman Road just before it enters the village of Leeming (postcode DL7

9SN). It has space for parking and is good view viewing and photographing movements, particularly off runway 34.

RAF Marham

Runways

01/19 6,115ft / 1,864m
06/24 9,131ft / 2,783m

Base for: No 138 Expeditionary Air Wing, No 207 Squadron, No 617 Squadron

Marham is one of the Royal Air Force's most active air bases, with a few based squadrons. Based aircraft include the very latest F-35B Lightnings.

The airfield is situated in Norfolk, close to Kings Lynn and Swaffham. It has two active runways, with operational areas all around the site.

Viewing locations are limited, but there are opportunities. If you drive along Chalk Lane, which follows the eastern perimeter, you'll pass the lights at the end of runway 24. Parking on the verge was permitted previously, but is not any longer, so you'll need to park elsewhere and walk here for good approach shots.

Similarly, the A1172 runs along the southern edge of the airfield and passes close to the end of runway 06. There is usually space to pull up at the side of the road, and other spotters will usually be doing the same. Here, shots of approaching aircraft are possible. Footpaths lead north through the fields from this spot which can take you closer to the runway threshold, and also give views across to the parking aprons.

RAF Mildenhall

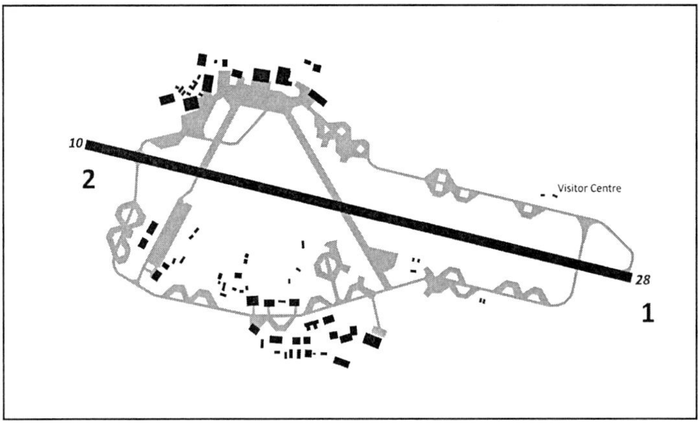

Runway

10/28 9,213ft / 2,808m

Base for: 100th Air Refuelling Wing, 352d Special Operations Wing, 501st Combat Support Wing

Mildenhall is a major military airfield at Bury St Edmunds in Suffolk, a short distance north east of Cambridge. Despite its RAF status, it is primarily used by the United States Air Force (USAF), and in particular their UK-based Air Refuelling Wing with based Boeing KC-135 Stratotankers.

Other based aircraft include CV-22 Ospreys and MC-130J Hercules. You'll also regularly see other USAF aircraft transiting through.

The large airfield has one active runway, with operational areas scattered both north and south of it.

SPOTTING LOCATIONS

1. Runway 28

A good spot exists for runway 28 arrivals and departures, which requires a little off road rambling. From Mildenhall town, head north along the A1101. After passing the petrol station on your left, go over the small roundabout, there's a wooded area on your left which you can walk through to reach a mound with elevated views above the perimeter fence. Parking is difficult along the road, so you may need to walk to the location from the industrial estate or Mildenhall town itself. Postcode for the place is IP28 7AL.

2. Nook Camp

A popular spot for watching and photographing movements is known as Nook Camp, with views of runway 10 arrivals and some of the parking ramps.

From Mildenhall town, head west along W Row Road, turning into Mildenhall Road, into Thistley Green. Turn north along Jarman's Lane and Pollards Lane, continuing to The Nook Campsite (postcode IP28 8RA). Go through the gate and follow the rough dirt track.

The spot is up against the fence, with an old trailer to help those without a ladder. You have great views.

Sometimes the site is closed in winter. There is an honesty box with a £1.50 per person suggested donation. You can use the toilet block in the caravan site.

RAF Northolt

Runways

07/25 5,525ft / 1,684m

Northolt is an active Royal Air Force base situated around 6 miles north of London Heathrow Airport. It is of interest to the civil enthusiast still due to its regular use by executive aircraft and the home of the Royal Flight aircraft.

Because of its sensitive nature and military use, Northolt is not the easiest of airports to spot at and extra care should be taken not to draw attention to yourself.

The airport has one main runway. To the south is the main parking apron for biz jets, whilst the military area is to the north. Most movements occur on weekdays.

West End Road/A4180 (postcode HA4 6NG) passes the end of runway 25. Just past the petrol station the hedge disappears and you have a view down the runway. Park somewhere nearby and walk to this spot and you'll be able to see most of the parked aircraft, albeit at a good distance.

Public Transport: Ruislip Gardens is the nearest London Underground station, and South Ruislip is the nearest mainline train station to RAF Northolt. Buses run past the base frequently.

RAF Shawbury

Runways

05/23 4,511ft / 1,375m
18/36 6,007ft / 1,831m

Base for: No 1 Flying Training School, No 60 Squadron, No 660 Squadron, No 670 Squadron

Shawbury is an air base just outside Shrewsbury in Shropshire. It has two active runways, but is mostly active with helicopter squadrons, including pilot training operations.

Shawbury also has a storage facility for RAF aircraft, which is unfortunately inside hangars.

A track runs past a farm on the northern side of the site which gives good views across the airfield. To reach it, head along the A49 and turn off for the tiny village of Acton Reynald (postcode SY4 4DS). Look for a right turn alongside a wall and follow it to the end.

Alternately, from the village of Shawbury itself, head south along the A53 towards Shrewsbury. Just before reaching a roundabout, turn right along a road which runs past the end of runway 36. From this general area you have some views across the airfield.

RAF Scampton

RAF Scampton in Lincolnshire is one of the most historic military bases in the county, but sadly closed in 2022.

Known as being the base of the Dambusters' 607 Squadron, it was also home in later years to the RAF Red Arrows aerobatic team and their hawk aircraft.

The Red Arrows have moved to RAF Waddington, and other based squadrons and personnel to other bases. The future of the site is not yet certain. However, it seems likely that some of the heritage structures and landmarks will be retained.

RAF Waddington

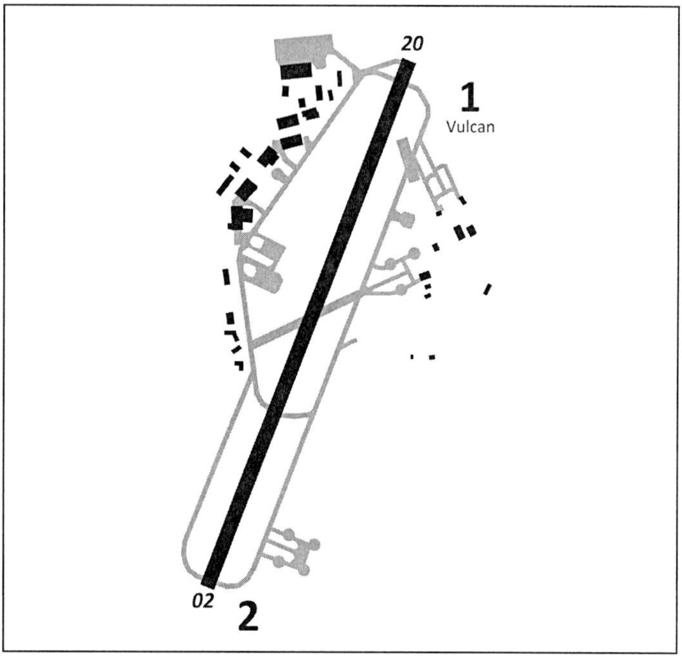

Runway

02/20 9,741ft / 2,969m

Base for: No 8 Squadron, No 13 Squadron, No 14 Squadron, No 51 Squadron, No 54 Squadron, No 56 Squadron, No 92 Squadron

Waddington is a busy air base just outside Lincoln in eastern England. It is home to the Intelligence Surveillance Target Acquisition and Reconnaissance (ISTAR) base, which uses RC-135W Rivet Joint aircraft – a derivative of the Boeing 707 – amongst others.

Traditionally Waddington was an Avro Vulcan base during the Cold War, and one example (XM607, which took part in the raids on the Falklands) is preserved here, usually close to the viewing area.

The airfield has one active runway. Most operational areas, aprons and hangars are on the western side, with the larger aircraft usually stationed outside their maintenance hangar at the northern side.

SPOTTING LOCATIONS

1. Wadding Aircraft Viewing Area

A dedicated viewing area exists on the north eastern side of the airfield, with good views across the runway. You can see most of the parking areas from here, and all movements.

The viewing area is well equipped with parking spaces and on-site toilet facilities and a café.

While the viewing area is good, spotters often wander across the road for better photographs at close range. However you'll need a ladder.

To reach the viewing area, head south from Lincoln on the A15 and you'll see it signposted on your left as you pass the runway 20 end.

2. Runway 02

At the opposite end of the airfield there are a couple of locations where you can go for photographing movements. You can also see some of the more remote parking areas here.

To reach it, continue south along the A15 from spot 1. Turn right along the B1178, signposted Harmston/Waddington. After a mile, look for a lay-by on the right which loops away from the road. You can park here, and walk along the track towards the runway end and crash gate.

RNAS Yeovilton

Runways

04/22 4,800ft / 1,463m
08/26 7,520ft / 2,292m

Base for: Fleet Air Arm

Alongside RNAS Culdrose, Yeovilton is the other significant Royal Naval Air Station in South West England. It is located at Yeovil in Somerset.

Yeovilton is home to various Wildcat and Merlin squadrons.

Operational areas are situated on the northern side of the main runway, with an apron on the south side used by the Fly Navy Heritage Trust and their historic aircraft (see navywings.org.uk)

Yeovilton is also home to the Fleet Air Arm Museum (see Museums section).

Spotting locations are scattered around the airfield perimeter. One good location is at the museum itself, which has a viewing gallery overlooking the main aircraft apron. You will need to purchase admission, however.

Not far from the museum, follow the road and turn right towards Bridgehampton, and then right again along Speckington Lane. This eventually runs to the end of runway 26, where you can observe and photograph movements.

Similarly, head west from the museum, past the base entrance, and turn right signposted Yeovilton Village/Fleet Air Arm Memorial Church. Where you see the runway you can park at the side of the road and observe movements easily.

Northern Ireland

Belfast City George Best Airport

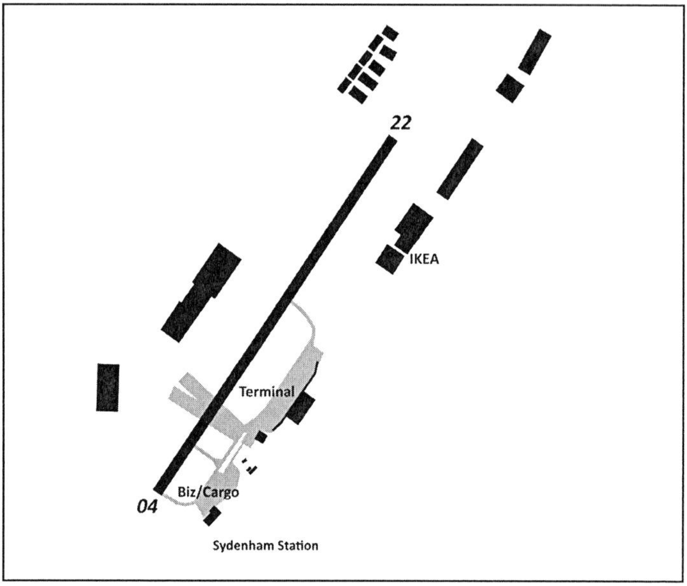

Runways

04/22 6,001ft / 1,829m

Hub For: Aer Lingus Regional, Flybe

Renamed after the local footballing legend, the former Belfast Harbour airport is a compact facility located within easy reach of the city centre and nearby docklands and Titanic attractions.

Traffic consists mostly of commuter flights to other UK cities. Other carriers include British Airways, easyJet, Eastern Airways, Loganair and KLM Cityhopper.

The passenger terminal is on the eastern side of the runway, close to the A2 motorway. Cargo and executive aircraft park on a separate ramp further south. Walking through the car parks from the terminal should yield views of aircraft parked here.

To the north of the terminal, alongside the runway, is a large IKEA store with a multi-storey car park which can be used for logging aircraft on the runway during store opening hours, providing you don't draw too much attention. It is accessed from the next exit of the A2 after the airport (follow signposts for Holywood Exchange). Photography is a bit limited due to trees.

Public Transport: Sydenham railway station is close to the airport terminal, linked from central Belfast. Buses are also regular.

Belfast International

Runways

07/25 9,121ft / 2,780m
17/35 6,204ft / 1,891m

Hub For: easyJet, Jet2, Ryanair, TUI Airways

Around 11 miles north west of Belfast is the main international airport for the city and Northern Ireland. It shares the site with a military helicopter base on the site of the former RAF Aldergrove which closed in 2008 and occupied the southern part of the airport.

The majority of passenger operations are by easyJet, Jet2, Ryanair, and TUI Airways. Scheduled services by mainline carriers tend to use Belfast City, and transatlantic links once enjoyed here have not yet returned since the pandemic. Virgin Atlantic provides seasonal services to Orlando.

Cargo flights are provided by DHL and some of its partners. It's easy to see these aircraft by walking left out of the passenger terminal. A private jet terminal and apron are located in the north-west corner of the airfield.

The airport has two runways running at right angles to each other. The aprons of the former RAF base are largely unused now. The cockpit of a Trident 2 airliner (G-AVFE) can be seen at the western extremity of the airport, used as a fire trainer.

A viewing gallery inside the passenger terminal was refurbished prior to the pandemic, but at the time of writing had not yet reopened. Check with the airport website before visiting. It offers views through glass of the airliner gates.

Fairly decent opportunities to photograph aircraft approaching runway 25 can be found about half a mile before touchdown. To reach the spot, drive east from the airport on Airport Road and turn right onto Drennans Road. You'll come to Killead Church, which has a car park that may be used if not obstructing anyone. Looking north, aircraft will pass close by on approach.

Public Transport: Translink's Airport Express 300 links Belfast International with the city centre and onwards to elsewhere in Northern Ireland.

City of Derry

Runways

08/26 6,460ft / 1,969m

A quiet airport in the north west of Northern Ireland around 7 miles east of Derry. It has a single active runway with two disused cross runways. The passenger terminal and parking apron is to the south of the runway, with flights provided by Loganair and Ryanair.

You can try following Clooney Road from the airport entrance, then turning right onto Airfield Road (postcode BT47 3PZ) towards the flying club where the road splits and runs to two different crash gates, where the runway can be seen.

On the north side of the airfield is another road leading to a crash gate close to a local equestrian centre (postcode BT47 3PZ).

Public Transport: Ulsterbus links the airport terminal with the city centre.

Scotland

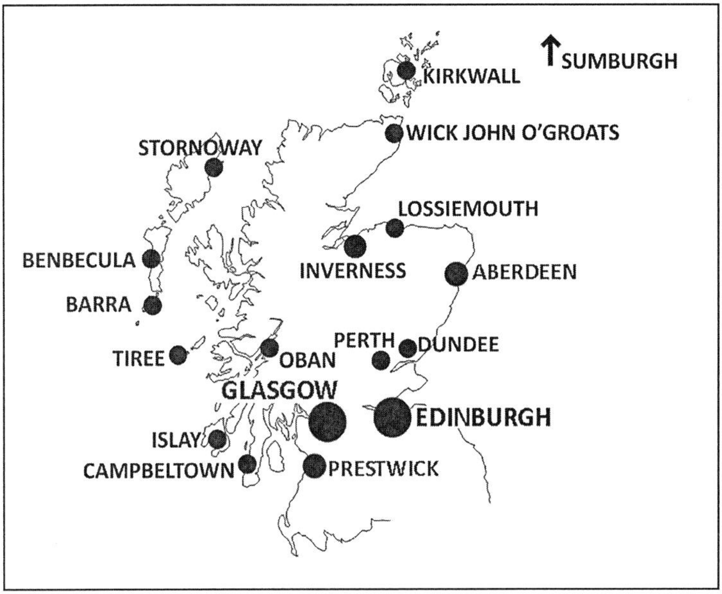
SUMBURGH
KIRKWALL
WICK JOHN O'GROATS
STORNOWAY
LOSSIEMOUTH
BENBECULA
INVERNESS
ABERDEEN
BARRA
PERTH
DUNDEE
TIREE
OBAN
GLASGOW
EDINBURGH
ISLAY
CAMPBELTOWN
PRESTWICK

Aberdeen

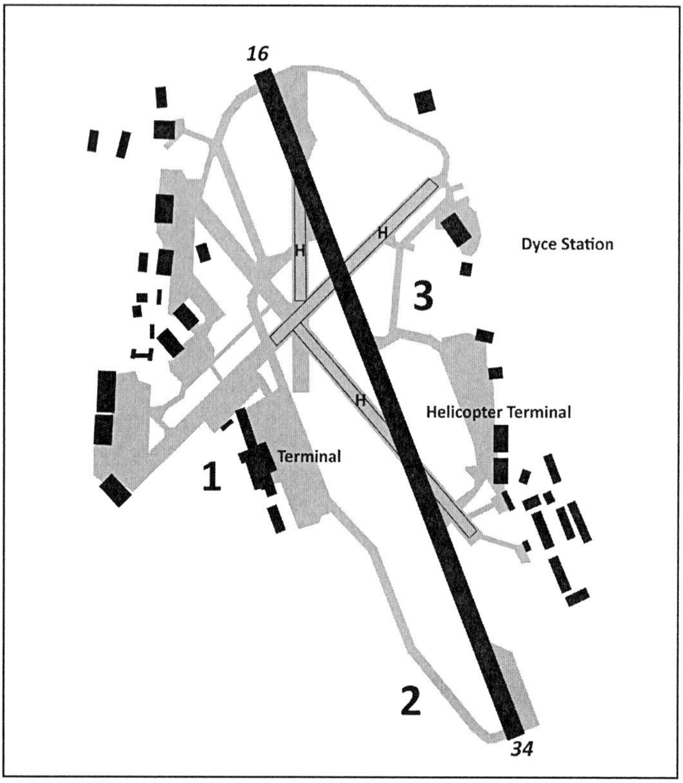

Runways

16/34 6,407ft / 1,953m

Hub For: Bristow Helicopters, CHC Helicopters, Loganair, TUI Airways

Aberdeen is a busy airport in north east Scotland which is a hub of activity surrounding the offshore oil industry, with much of its traffic made up from helicopters ferrying workers to and from rigs.

Airline flights to Aberdeen are often related to this also, with links to most regional airports across the UK and many Scandinavian airports. Some seasonal flights exist to holiday destinations, with TUI operating based aircraft, and cargo flights are operated.

The passenger terminal is to the west of the runway, with a dedicated helicopter terminal on the east side in Dyce village. Helicopter operators have maintenance bases to the north of the passenger terminal

SPOTTING LOCATIONS

1. Terminal Car Park

The top of the car park outside the passenger terminal is good for quick views of aircraft taxiing and parked on some of the ramps, including the helicopters. Photography is possible, but you may be moved on from this location.

2. Runway 34

From the Budget Car Hire building to the south of the passenger terminal, a disused road and track runs along the perimeter fence. Following it takes you to a spot opposite the runway 34 threshold, where you can spot and photograph through the fence. A viewing shelter used to exist here but has been removed.

3. Cordyce View

On the eastern side of the airport Cordyce View (postcode AB21 7DS) is a residential street which runs to a crash gate. Opposite the houses is the fence next to the helicopter terminal and executive aircraft apron, which you can photograph through. The crash gate also looks out onto the runway, and has a small space for parking cars. Alternatively you can walk 5 minutes from nearby Dyce railway station.

From Cordyce View, you can walk along Foinavon Close, where there is an adjacent grass mound overlooking the airport from an elevated position, which is better for photographs.

HOTELS

Crowne Plaza Aberdeen Airport

2 International Gate, Dyce, Aberdeen AB21 0BE | +44 1224 608350 | *www.ihg.com*

This hotel is just to the south of the terminal, close to other hotels but it seems it has the best views of all of them (along with the Holiday Inn Express). Ask for a room facing the runway for views of aircraft movements and helicopter operations.

Public Transport: Bus #727 links the airport to the city, whilst Dyce railway station links the city to the helicopter terminal. Bus #80 travels between the passenger terminal and helicopter terminal.

Barra

Runways

07/25 2,621ft / 799m
11/29 2,231ft / 680m
15/33 2,776ft / 846m

Probably the most unusual airport in the British Isles. Barra's three runways are all on the beach. Daily Loganair Twin Otter services link Barra with Glasgow, with arrival times governed by the times of the tides. Aircraft parking up next to the sand dunes and small terminal building. Good views of aircraft movements can be had from the area north of the terminal.

Benbecula

Runways

06/24 6,023ft / 1,836m
17/35 4,003ft / 1,220m

A rural airport with two runways on the island of the same name in the Outer Hebrides. Daily services to Glasgow and Stornway with Loganair. The passenger terminal and parking apron are at the southern side of the airport, with good views through the fence. The access road also has views of the runway.

Camplbeltown

Runways

11/29 5,741ft / 1,750m

Despite its rural island location, Campbeltown has the longest runway in Scotland, hinting at its former role as a NATO base. Today it is sustained by a public service flight link to Glasgow with Loganair DHC-6 Twin Otters. The terminal is in the eastern part of the site.

Public Transport: Bus #443 links the airport to Cambeltown.

Dundee

Runways

09/27 4,593ft / 1,400m

A small regional airport on the banks of the River Tay, which has had mixed fortunes with commercial services. Loganair provides a few daily scheduled turboprop flights, alongside a busy general aviation community and occasional executive aircraft movements. The passenger terminal and GA parking are at the eastern end of the airport. Views are possible from the flying club car parks.

Public Transport: Bus #5 and #9 pass close to the airport from the city centre and railway station.

Edinburgh

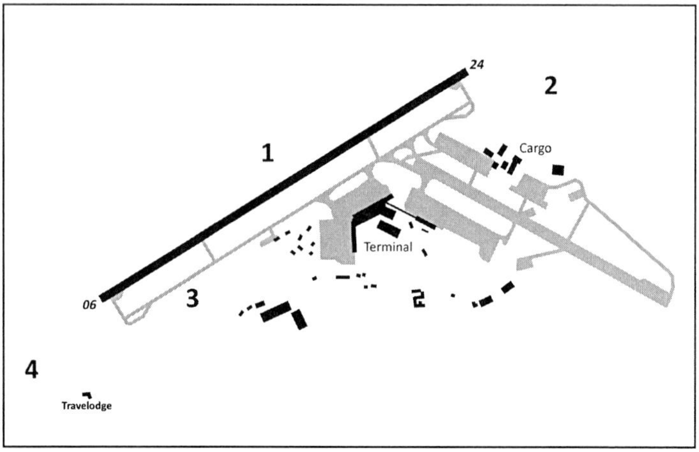

Runways

06/24 8,386ft / 2,556m

Hub For: easyJet, Jet2, Loganair, Ryanair

Scotland's busiest airport. Edinburgh Airport has grown significantly in recent years, managing to pull off expansion in full service, low cost and cargo services. Key operators of interest to the enthusiast include Air Canada, Atlantic Airways, Delta, Finnair, Lufthansa, Norwegian Air Shuttle, Qatar Airways, Scandinavian Airlines, Turkish Airlines, United Airlines, Vueling and Wizz Air.

The passenger terminal occupies a central position on the airport site, with the runway to the north. The former secondary runway is now closed, earmarked for more parking areas; the cargo base is to its east.

SPOTTING LOCATIONS

1. River Almond Bridge Gate

On the northern perimeter of the airport at roughly the mid-point of the main runway is a crash gate just off a bridge over the River Almond, which is popular with local spotters. There is room for a few cars to park and you can see and photograph all movements on the runway. To reach the location head to the village of Kirkliston (beyond the airport, off the M9 motorway). At the village centre head east on Main Street (postcode EH29 9AD) for a mile or so.

2. Lennymuir Railway Bridge

Turnhouse Road (postcode EH12 0AZ) runs to the cargo site on the east of the airport. You can't get any good views of these aircraft, but if you park your car and walk along Lennymuir (EH12 0AP) there is a railway bridge with good photographic opportunities of aircraft approaching runway 24.

3. Western Car Parks

A 15 minute walk west of the passenger terminal is a large open air car park which fronts the perimeter fence alongside the taxiway and runway 06 threshold. Head for the Holiday Inn Express hotel and continue past it.

4. Glasgow Road Park

As the A8 (Glasgow Road) passes to the west of the airport, a small park with benches by the side of the road (opposite the Travelodge hotel) has a great view of aircraft just before touchdown on runway 06.

HOTELS

Travelodge Edinburgh Airport

Ratho Park, Glasgow Rd, Edinburgh EH28 8PP | 0871 984 6340 | *www.travelodge.com*

Ask for a room overlooking the A8/Glasgow Road. These face the threshold of runway 06, giving you views of most movements. It would be beneficial to be on the top floor as trees outside the hotel can obscure the view.

Public Transport: Trams link Edinburgh Airport with the city centre, alongside buses.

Glasgow International

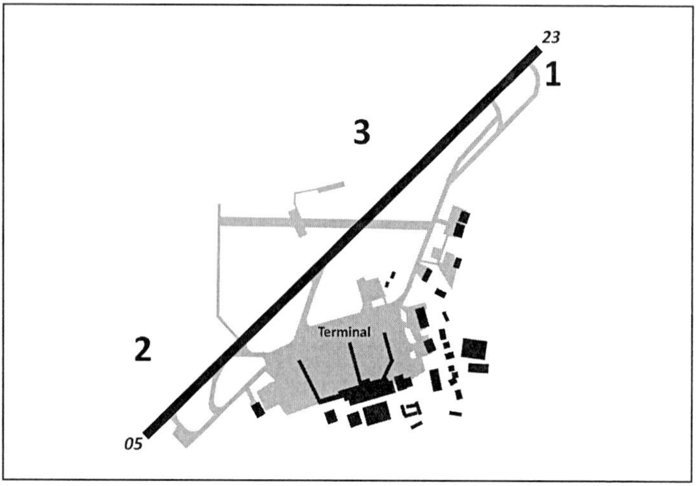

Runways

05/23 8,743ft / 2,665m

Hub For: easyJet, Jet2, Loganair, TUI Airways

Glasgow's main airport is a busy gateway to Scotland, with many domestic, European and long haul links. Many long-haul routes are seasonal, such as those by Air Transat and WestJet. However, Emirates provides year-round service, and there are many low-cost and scheduled services to keep spotters interested, as well as the interesting island-hopping services to the remoter parts of Scotland.

The passenger terminal sits to the south of the runway and has three piers. There are no real views from the terminal unless you have a flight ticket, so it's best to head to one of the locations around the perimeter.

SPOTTING LOCATIONS

1. Abbotsinch Road & Runway 23

From the terminal head under the M8 and then turn left onto Abbotsinch Road. This passes the Loganair hangars, cargo area and FBOs on the eastern side, so it's worth exploring before continuing along the road until you find a layby on the left which looks out to the threshold of runway 23. Aircraft can be photographed landing but you may wish to walk back along the road to find a better view of aircraft on the ground.

2. Barnsford Road

From the terminal head west for Barnsford Road/A726 (avoid joining the M6 motorway) which loops around the end of runway 05. A crash gate adjacent to the threshold is popular, but there's only space for one car to park. A footpath along the road helps you find a good spot to view and photograph through the fence.

3. Walkinshaw Road

Continuing along Barnsford Road you'll come to a small lane called Walkinshaw Road on your right (postcode PA4 9LP) which runs along the northern perimeter to a farm and crash gate. Do not block the road or crash gate, or park on the farm's land. You have a good view across the airport and runway here, with most of the terminal in view and decent photography opportunities through the fence.

Public Transport: Bus #500 links the airport with Glasgow city centre, and #757 with Paisley.

Glasgow Prestwick

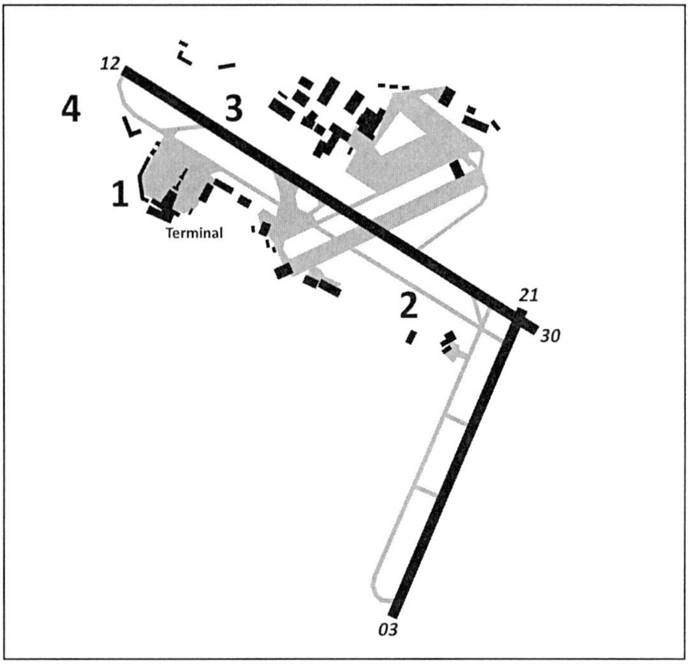

Runways

12/30 9,797ft / 2,986m
03/21 6,250ft / 1,905m

Hub For: Cargolux, Ryanair

Prestwick in Ayrshire is no longer the stopover point for aircraft heading across the Atlantic, but it does still see some interesting traffic – particularly military and cargo movements. Passenger flights are now limited to Ryanair, however freight services by the likes of Air France Cargo and Cargolux are common.

Prestwick has two runways at right angles. The passenger terminal and cargo apron are at the western side, with hangars and maintenance areas on the north side.

SPOTTING LOCATIONS

1. Terminal Viewing Gallery

You can view and photograph airliners on the ground from the spectator's galley upstairs in the passenger terminal. Photography is through glass.

2. Spotting Mound

A popular unofficial place to watch movements is a mound and area of waste ground close to the intersection of the two runways. It has views across to the apron used by biz jets and some cargo aircraft, and excellent views of the runway. To reach the location head east from the terminal along the A79, Shawfarm Road, and then left onto Shaw Road (postcode KA9 2LN). The track to the waste ground is on your left.

3. Monkton

On the northern side of the airport in the village of Monkton follow Main Street (postcode KA9 2QL) south where you'll come to the fence. Views through the fence are of aircraft lining up and approaching runway 12, albeit mostly into the sun.

4. Railway Bridge

Just to the west of the terminal, off the A79, is a road leading to the Prestwick Holiday Park (signposted). The road crosses a railway line on an bridge, which is a great spot for photographing runway 12 arrivals. There's usually spots to park your car.

Public Transport: A railway station outside the terminal gives swift access to central Glasgow, Ayr, Kilmarnock and Paisley. Bus #4 links the airport with Monkton, with a short walk to location 3.

Inverness

Runways

05/23 6,191ft / 1,887m
12/30 2,297ft / 700m

Hub For: Loganair

The capital of the Highlands. Inverness Airport is a few miles east of the city at Dalcross and is a hub for flights around the country and to the Orkney and Shetland islands.

Loganair is the main operator, with cargo flights and passenger flights. Other carriers include British Airways, easyJet and KLM Cityhopper, and seasonal holiday flights.

A popular spotting location can be found just inside the airport entrance (postcode IV2 7JB). At the second roundabout, turn right down a dead end which leads to a crash gate. Spotters often park here, and you have a good view of arrivals on runway 05 and can see any aircraft parked at the terminal.

North of the terminal, the area around the Highland Aviation and Signature Flight Support buildings have limited views of the general aviation ramp and any biz jets parked up.

Public Transport: Bus routes link Inverness Airport with both Nain and Inverness city centre.

Islay

Runways

08/26 2,083ft / 635m
13/31 5,069ft / 1,545m

In the southern part of the island around 5 miles north of Port Ellen. Regular scheduled services by Loganair to Glasgow, and other communities with Hebridean Air Services Islanders. Any parked aircraft can be seen from the passenger terminal.

Public Transport: Bus #451 links the airport with Port Ellen.

Kirkwall

Runways

09/27 4,685ft / 1,428m
14/32 2,231ft / 680m

The principal gateway to the Orkney Islands at the main town. It is served by Loganair services to Scotland's main airports, and local Loganair services to remote airports across the Orkneys and Shetlands. Aircraft can be seen from around the terminal area, or the separate GA area to the west.

Public Transport: Bus #4 runs between the airport and the town/ harbour.

Oban

Runways

01/19 4,147ft / 1,264m

A scenic small airport on the west coast of Scotland and around 6 miles north of the town which has rail and ferry links. The airport is only served by local Hebridean Air Services links using Islander aircraft, but is a popular spot for visiting GA aircraft.

You can see parked aircraft easily from the passenger terminal. There is also a crash gate looking out onto the southern runway boundary on Lora View (postcode PA37 1RR).

Public Transport: Buses #005 and #918 both link Oban with the airport terminal.

Perth

Runways

03/21 2,799ft / 853m
09/27 1,998ft / 609m
15/33 2,034ft / 620m

Perth is not a commercial airport, but is one of the busiest general aviation airfields in Scotland. It is three miles northeast of the city and has a number of flight training schools based.

If you bring a hi-viz vest and ask at the tower or flying schools you are likely to be permitted access airside to see aircraft on the ground and in the hangars.

Public Transport: Bus #57 links Perth Airport with Perth and Scone.

Stornoway

Runways

06/24 3,281ft / 1,000m
18/36 7,595ft / 2,315m

One of the busier airports in the Scottish islands. Stornoway is served by Loganair on flights around the country. It also handles mail flights and has a small helicopter base for the Coastguard.

The passenger terminal is on the eastern side of the airport and aircraft can easily be seen through the fence.

Public Transport: The #W5 bus service links Stornoway town centre and ferry port with the airport.

Sumburgh

Runways

09/27 4,921ft / 1,500m
15/33 4,678ft / 1,426m

The main airport for the Shetland Islands, situated at the southern tip and some distance from the main town at Lerwick, which has its own tiny airport.

Sumburg has two main runways and a small helicopter runway. The main 09/27 strip has a road running across the 09 threshold with barriers to stop traffic when movements occur. Movements are mainly by Loganair and Bristow Helicopters.

The passenger and helicopter terminal is on the airport's eastern side, meaning the access road runs a circuitous journey around the airport perimeter, giving ample views of the runways and parked aircraft. Walls around the car park unfortunately limit views of parked aircraft, but the gravel parking area on the entrance road near the Sumburgh Airport Memorial has good views of any movements.

Public Transport: A bus service links Sumburgh Airport with Lerwick.

Tiree

Runways

05/23 4,829ft / 1,472m
11/29 2,621ft / 799m
17/35 2,598ft / 792m

A former RAF training base on the island of Tiree. It has a small terminal building and parking apron on the west side of the airport, with daily Loganair flights to Glasgow. Aircraft parked on the apron are visible from the terminal area.

Wick John O'Groats

Runways

13/31 5,988ft / 1,825m

Wick is a small airport that has recently been rebranded to attract travellers to the nearby John O'Groats at the northeast corner of the British mainland. Nevertheless, the airport remains quiet, served only by Eastern Airways from a small terminal which has views of any aircraft parked up.

The airport does see a good number of transiting aircraft making the journey across the Atlantic, which might yield an interesting sighting.

The hangars and business park limit views from the entrance road.

Scotland's Military Airfields

RAF Lossiemouth

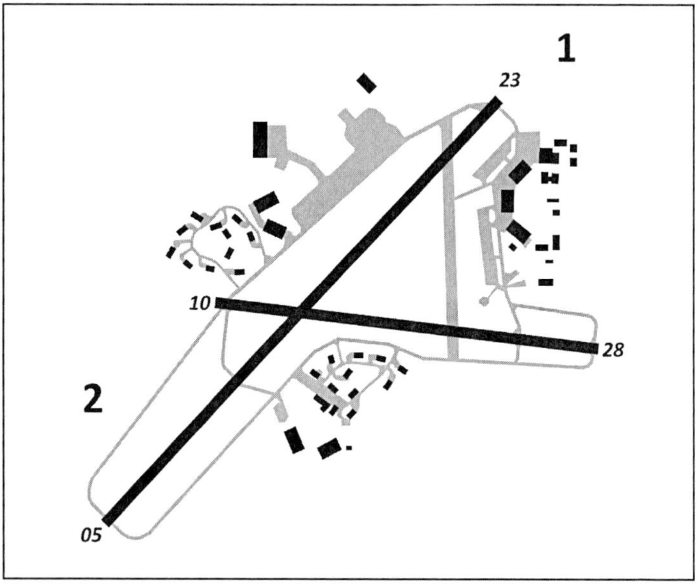

Runways

05/23 9,068ft / 2,764m
10/28 6,070ft / 1,850m

Base for: No 1 Squadron, No 2 Squadron, No 6 Squadron, No 9 Squadron, No 120 Squadron, No 201 Squadron

The only remaining operational Royal Air Force base in Scotland, following the closure of RAF Lecuhars in 2015. Lossiemouth is one of the most active fast-jet stations in the country, and situated right on the coast, which makes for excellent training areas and good proximity to any threats from the east.

Today Lossiemouth is home to Typhoon fast jets, and Boeing 737-based Poseidon MRA1 maritime patrol and reconnaissance aircraft. E-7 Wedgetail aircraft are also expected to be positioned there.

The operational areas are on either side of runway 05/23

SPOTTING LOCATIONS

1. Golf Course

Park your car in the small gravel car park near the entrance to the base, where the B9040 and B9135 meet. Then walk back along the road into the golf course. A path on your left leads to the perimeter fence, which you can follow along toward the end of runway 23. An elevated spot has views, and approaching aircraft can be photographed.

2. Runway 05

From the entrance road through the base (B9135), keep going south, past the end of runway 28, until you come to a road on your right (Drainie Road). Follow it as it curves around the southern perimeter. Just after you see the end of runway 05 on your right, turn right along a narrow road, where you'll be able to find a good spot to photograph arrivals. There is some roadside parking.

Wales

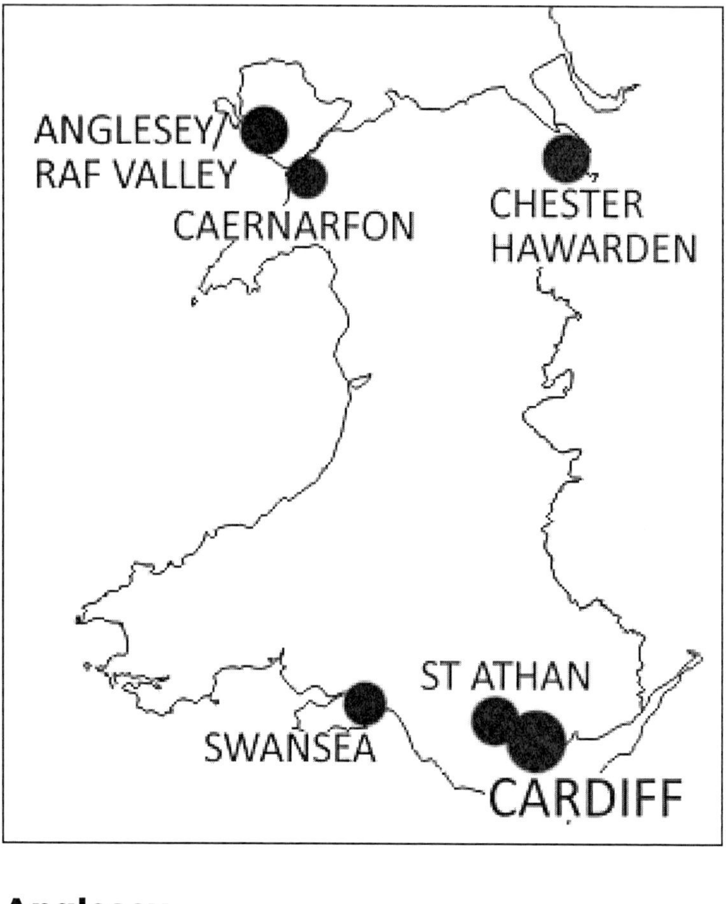

Anglesey

Runways

01/19 5,377ft / 1,639m
08/26 4,200ft / 1,280m
14/32 7,513ft / 2,290m

Anglesey Airport occupies part of the RAF Valley base, which is an active training centre for the Royal Air Force's Hawk aircraft. The airport had a subsidised passenger air service linking it to Cardiff until recently, but at present has no commercial flights.

A large parking area alongside Minffordd Road (postcode LL65 3NA) at the entrance to the passenger terminal and base is a great place to watch movements. There is a Hawk jet gate guardian alongside and a good view of the runway 19 approach and nearby parking aprons.

Public Transport: Bus #X4 links the airfield with central Holyhead. It passes the spotting location mentioned above.

Caernarfon

Runways

02/20 3,543ft / 1,080m
07/25 3,058ft / 932m

A sometimes busy general aviation airfield on the north west coast of Wales, near the southern tip of Anglesey. It is home to a helicopter search and rescue base, and also a small aviation museum and visitor centre (see www.airworldmuseum.com). Views possible from the main car park, as well as the nearby beach and car park.

Cardiff

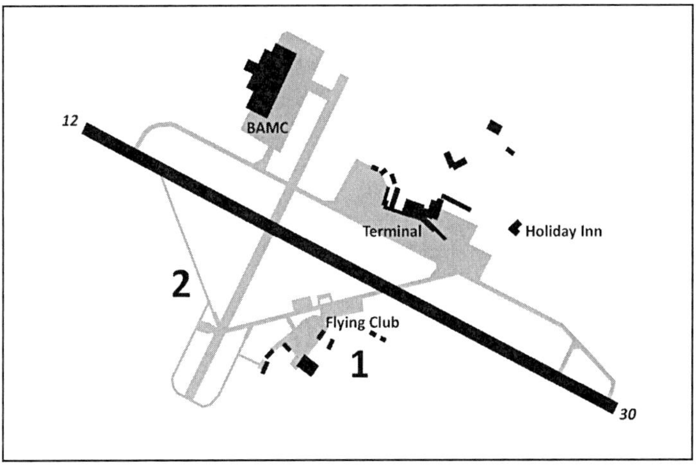

Runways

12/30 7,723ft / 2,354m

Hub For: Wizz Air UK

The busiest and largest airport in Wales, serving its capital on the south coast. Cardiff handles over a million passengers per year, mostly on regional flights and holiday charters, operated by airlines like TUI, Vueling and Wizz Air. Other carriers include Aer Lingus Regional, KLM Cityhopper and Loganair.

The airport is home to the British Airways Maintenance Cardiff site (BAMC), which occupies a large hangar and separate aircraft apron to the north west of the passenger terminal. Here the airline's long haul fleet are maintained. This gives the airport an interesting mix of visiting aircraft, but it is nothing you won't see in abundance at Heathrow.

The south side of the runway is home to a flying club and FBO for private aircraft. Here you'll find the Cambrian Flying Bar (1), which is an excellent viewing location with an outdoor balcony and views across to the runway and terminal. Reach it by following Porthkerry Road around the end of runway 30 and following signs (postcode CF62 3EQ). You can also see some of the parked aircraft from the car park outside the club.

Another popular viewing location is Port Road (postcode CF62 3BH), beyond the Highwayman Inn, which has a track running along the fenceline with views of the runway (2). You may need a ladder for photographs. Look out for the BAC One-Eleven fire trainer fuselage (G-AVMT) alongside the former cross runway nearby.

Public Transport: Cardiff Airport is linked to the city via the TrawsCymru T9 bus every 20 minutes. Bus #303 links the airport with Rhoose, with the Station Road stop a short walk from the flying club.

Hawarden

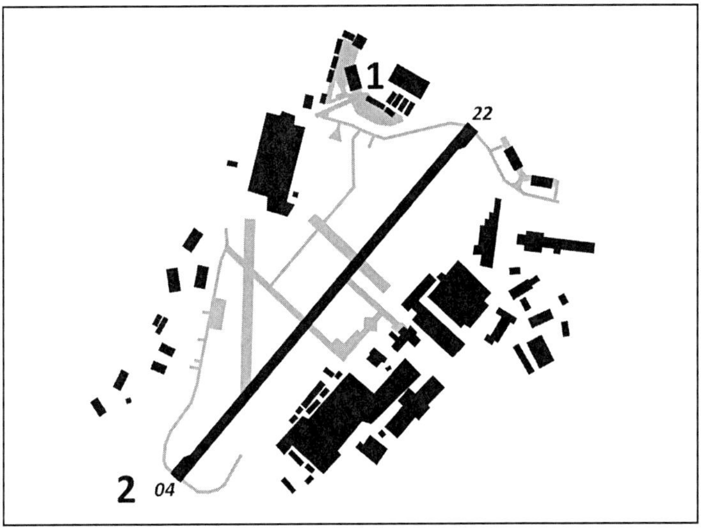

Runways

04/22 6,699ft / 2,042m

Hub For: Airbus

Built close to the English border and the city of Chester, Hawarden is a Welsh airport heavily involved in the production of parts for Airbus aircraft. These are transported by giant Beluga transport aircraft to the factories in Hamburg and Toulouse most days of the week.

Although no passenger services exist, airlines have provided employee flights from Hawarden to Airbus sites in Bristol and Toulouse in the past.

Hawarden is also a good place to catch biz jets either involved in Airbus activities, or visiting the on-site Marshall Aerospace servicing facility which looks after Beechcraft and Cessna types.

Security can be intolerant of spotters and ask them to leave if they find you peering or photographing through the fence.

SPOTTING LOCATIONS

1. Chocks Away Diner

The Chocks Away Diner (www.chocksawaydiner.co.uk) at the general aviation area on the north side of the airfield is a good place to eat and watch aircraft movements. Open daily.

2. Runway 04 End

The B5125 Chester Road (head for postcode CH4 0NQ, St Mary's Way) runs past the end of runway 04 with a great view along its length. With a good pair of binoculars you'll see most parts of the airfield. You cannot park here, so find a place in nearby Broughton and walk a few minutes to find a good spot.

Public Transport: Bus #4 and #21 from central Chester to Hawarden pass along Chester Road on their way to the depot on the west side of the airport.

MOD St Athan

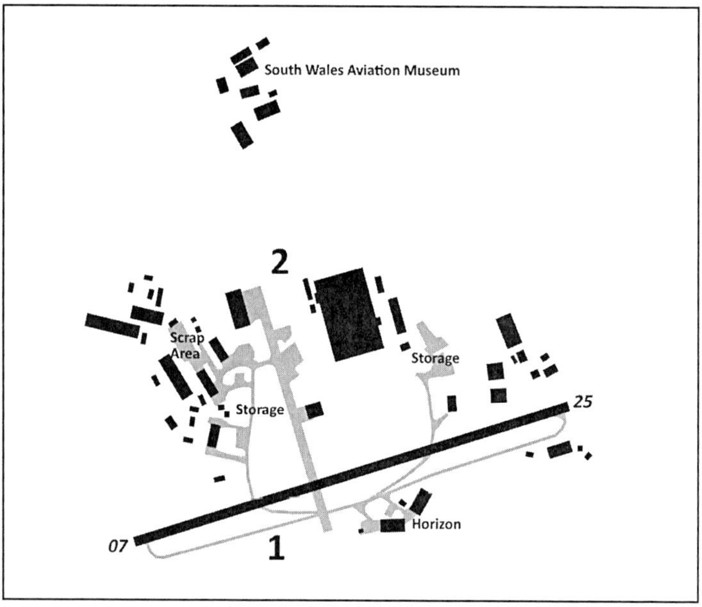

Runways

07/25 5,997ft / 1,828m

St Athan is a few miles west of Cardiff Airport. Part of the airfield is still an active Ministry of Defence site which has been heavily involved in Royal Air Force training in the past, but does not see military aircraft movements. The airfield is a great draw as one of the UK's main airliner storage and scrapping locations. It also has an interesting aviation museum to visit.

All of the airfield's buildings, hangars and aprons are on the north side of the runway. Movements are few, but a drive around should uncover any stored airliners present.

From St Athan village, drive west along Llantwit Road. You'll come to the hangar of Horizon Training on your right, and just after is a view through the fence of anything they have parked outside. Carry on along the road and you'll come to a crash gate with room to park (1). From here you can see across the runway to part of the storage areas.

Once completed here, head back east along Llantwit Road, and turn right after the Horizon hangar along a track which crosses a railway bridge. Turn right onto the B4265 and follow it until you come to a roundabout. Turn right (second exit) onto Northern Access Road which will lead to an area with some views through the fence of the scrapping and storage aprons on your right. Take the usual precautions not to obstruct and gates or access areas.

The South Wales Aviation Museum is just north of this spot (see Museums section).

Public Transport: Bus #304 links the northern part of St Athan Airport with Rhoose, near Cardiff Airport.

Swansea

Runways

04/22 4,429ft / 1,350m
10/28 2,812ft / 857m

Efforts to bring commercial services to Swansea have never gone far, with the last flights ending in 2004. It is mainly used for general aviation, training and helicopter rescue flights, but it is hoped to develop the airport with new facilities to attract more aviation business. The car park outside the flying club is a good place to log any aircraft on the ground. It has views through the fence to the main runway.

Public Transport: Bus #118 from central Swansea passes Swansea Airport.

Channel Islands and Other Islands

Alderney

Runways

03G/21G 1,631ft / 597m
08/26 2,877ft / 877m
13G/31G 2,405ft / 733m

Hub For: Air Alderney

A small airport on the tiny island of Alderney. Of its three runways in a triangle pattern, only 08/26 has a hard surface. The small terminal building is due to be replaced or upgraded, but the airport is unlikely to get any busier. Links to Guernsey and Southampton operated by Aurigny, and regular flights by Air Alderney. You can usually see parked aircraft from the terminal, or from surrounding roads.

Bembridge, Isle of Wight

Runways

12/30 2,713ft / 827m

Bembridge is one of two airfields on the Isle of Wight. It does not have any passenger services, but is popular among the GA community. It is home to Britten-Norman, whose aircraft are now built elsewhere but sometimes present. You can log aircraft from outside the tower.

Public Transport: Bus #8 passes Bembridge Airport from both Newport and Bembridge.

Guernsey

Runways

09/27 4,800ft / 1,463m

Hub For: Aurigny

The second busiest of the Channel Islands. Guernsey is a hub for intra-island flights, with Aurigny based here. Other airlines of note are Blue Islands and British Airways, and often summer charter series are flown from Germany. Biz jets are also common here.

The views in the terminal are usually sufficient for watching movements and logging airliners. You can also use the car park on the north side of the runway off La Villiaze Road which has views across the airfield. If you don't have a car, it is a 25-minute walk from the terminal.

Guernsey has a number of wrecked aircraft on the fire dump north of the runway. Approaching the fire service may yield a visit.

Public Transport: CT Plus operates three different bus services to Guernsey Airport, one of which loops the island.

Isle of Man

Runways

03/21 4,117ft / 1,255m
08/26 6,923ft / 2,110m

Commonly known as Ronaldsway Airport. Situated at the southern end of the island between Castletown and Ballasalla. Most traffic is by regional turboprop aircraft from the likes of Aer Lingus Regional and Loganair. easyJet and British Airways bring jet aircraft, but the runway lengths limit anything large.

There is an area to view aircraft movements in the terminal which overlooks the parking ramp and runways beyond. Alternatively, Derbyhaven is a small village on the coast to the south of the airport which has views across the airfield. Follow the A5 to Castletown and then Derbyhaven road east from the Promenade to reach the area. You can walk here in 30 minutes, but there is no public transport.

Café Bar Two Six is a little further along Airport Road from Derbyhaven. It is a pleasant little restaurant and café, sat next to the threshold of runway 26, with good views of movements.

The Manx Aviation and Military Museum has some aircraft exhibits, just south of the terminal. See maps.org.im.

Isles of Scilly St. Mary's Airport

Runways

09/27 1,722ft / 525m
14/32 2,277ft / 694m

Hub For: Isles of Scilly Skybus

St. Mary's is a small two-runway hilltop airport which handles flights by Isles of Scilly Skybus on most days to Lands End, Newquay and Exeter. Some GA aircraft are also based here. Aircraft can be seen from the terminal or by following High Cross Lane around the northern end of runway 14.

Isle of Wight Sandown

Runways

05/23 2,900ft / 884m

A very popular general aviation airfield on the Isle of Wight. Particularly busy during the summer months and weekends. All movements can be seen from the café area. You'll find the Wight Aviation Museum on site, too.

Jersey

Runways

08/26 5,597ft / 1,706m

Hub For: Blue Islands

The busiest of the Channel Island airports, surviving on a steady mix of scheduled traffic to both France and the British mainland, and a range of summer charters. Because links with Germany are still strong it is common to see summer flights from Eurowings and Lufthansa. Principal carriers at Jersey are Aer Lingus Regional, Blue Islands, British Airways, easyJet and Jet2.

The large, modern terminal does not have any viewing areas unless airside. To the east the road passes the biz jet parking apron and reaches the grass GA parking area. You can see through the fence here, but at the Jersey Aero Club there is a balcony and café/restaurant which is perfect for watching aircraft movements.

If you have a car there are two positions on the northern perimeter where you can park and watch movements through the fence. The first is on La Route de l'Hermite just outside St. Peter's. The other is along Rue de l'Eglise heading west out of St. Peter's, with a handy footpath alongside the fence to find the best spot.

Public Transport: Bus #15 links St. Helier to Jersey Airport.

Republic of Ireland

Cork

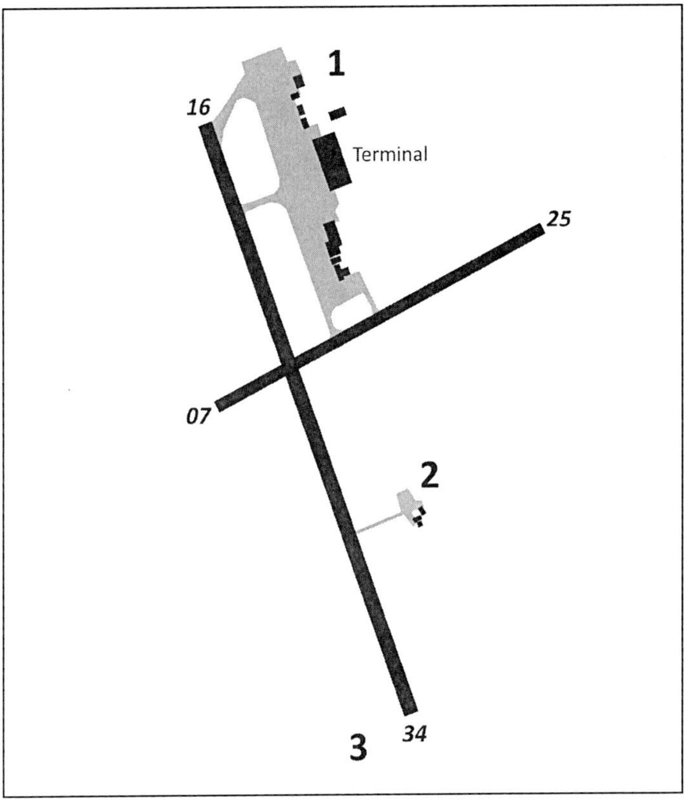

Runways

07/25 4,298ft / 1,310m
16/34 6,998ft / 2,133m

Hub For: Aer Lingus, Aer Lingus Regional, Ryanair

Less than 5 miles south of the city, Cork Airport is among the Republic of Ireland's busier airports. It is a fairly small airport, with two runways crossing each other. The terminal is in the north east corner, alongside a small cargo terminal and apron. Light aircraft and helicopters use a separate parking area linked to the southern end of the longest runway, 16/34.

Both main Irish carriers, Aer Lingus and Ryanair, maintain a heavy presence here with a mix of regional and leisure routes. Some other European carriers offer seasonal links to Cork, but the variety is not extensive.

The top floor of the multi-storey car park next to the terminal (1) is a good place to see aircraft on the ground and using the main runway.

You can also drive to the flying school by following R600 south (signposted Kinsale from the terminal access road) and turning into their car park (2). From here you can see any aircraft parked on their ramp, and also the main runway beyond.

Finally, a popular spot for watching aircraft movements particularly when arriving and departing on runway 34 is a layby at Lios Cross (3). Follow the R600 south and turn right onto the small road in Bowen's Cross. Follow this for a mile and you'll see the layby and airport.

Public Transport: Bus routes #225 and #226 link Cork Airport with the city centre every hour.

Donegal

Runways

03/21 4,908ft / 1,496m

A remote and scenic airport in the extreme North West of Ireland with a single runway between a beach and tidal area.

The airport has a small parking apron and terminal at the southern end, with views from the car park and the road passing the southern end of the runway. The only airline movements are ATRs from Aer Lingus Regional, and regular helicopter services to offshore rigs.

Dublin

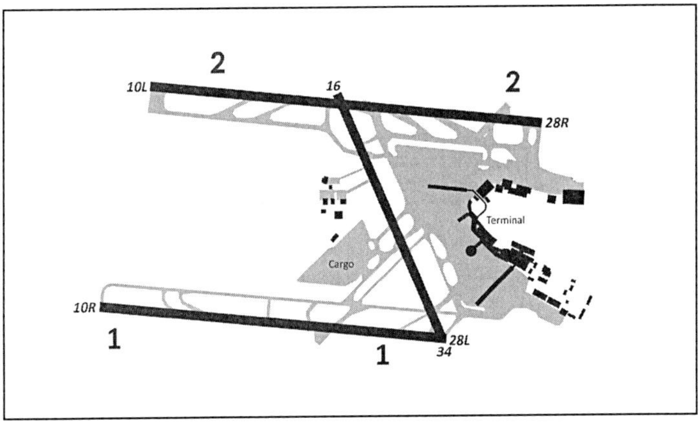

Runways

10L/28R 10,203ft / 3,110m
10R/28L 8,652ft / 2,637m
16/34 6,798ft / 2,072m

Hub For: Aer Lingus, Aer Lingus Regional, ASL Airlines Ireland, Ryanair, TUI Airways

When built in 1940, the terminal at Dublin's airport was considered to be the best in Europe and won architectural accolades for its designer. Following the Second World War the number of services increased rapidly. Three hard runways were laid, and new hangars and other buildings built. A terminal extension could only cope a few years into the long-haul jet revolution, and in 1972 another new building was opened.

The passenger terminal has seen steady expansion since the early 2000s. Many European, Middle Eastern and North American carriers have a regular presence here, alongside the based airlines.

Most recently, in 2022, a new parallel runway was opened along the northern part of the site, which has seen some of the local roads and old viewing points disappear. Runway 16/34 is used when wind conditions require, and is often used as a taxiway.

Dublin has become one of the fastest growing European airports and the US pre-clearance facility has propelled the stature of Aer

Lingus' transatlantic services, combined with connecting regional flights from across the UK and Europe.

Biz jets park on an apron at the northern side of the airport, which is difficult to see from the terminal or even aircraft windows. The cargo apron is opposite the passenger terminal, to the north of the main runway. Look out for a retired Ryanair Boeing 737-200 (EI-CJD) in the same area.

Sadly no official spotting facilities remain at Dublin, but a number of locations around the perimeter are popular.

SPOTTING LOCATIONS

1. Runway 10R/28L

Following the road away from the terminal, turn on to Old Airport Road, which follows the perimeter. You'll soon be alongside the southern runway, and depending on the direction in use you can find raised spots at either end of the runway which give you an elevated position for photography. Many spotters congregate here. Photography is good, and most traffic will pass you eventually. Follow the road around for spots on the northern side of the runway. Buses #16 and #41 heading for Dublin will drop you ½ a mile from this location (Dardistown Cemetery). A car is recommended.

2. Northern Perimeter

The new runway 10L/28R has re-routed the roads along the northern perimeter. Road R108, which becomes Naul Road further east, runs the length of the new runway and there are two lay-bys constructed alongside which have parking spaces and relatively good views across to the runway. One lay-by is at the eastern end, and the other the western end. Photography may be limited due to the perimeter fence and passing traffic.

To reach them, head away from the passenger terminal and follow signs for R132/Swords. At the next roundabout, take the first exit (signposted M1 (M50) Dublin and Belfast). The first lay-by is on your right after around a mile, but has a large embankment blocking part of the view. To reach the second lay-by, which is more open, continue and take the first exit at the next roundabout. It is on your right shortly after.

Radisson Blu Dublin Airport

Dublin Airport | +353 1 844 6000 | *www.radissonblu.com*

Rooms facing the airport have a view of arrivals on runway 28 and 34, and distant views of departures. You can see parts of the international pier, but no other gate areas. Too distant for photography.

Carlton Hotel Dublin Airport

Old Airport Rd, Cloghran, Dublin, K67 P5C7, Ireland |
+353 1 866 7500 | *www.carltondublinairport.com*

Situated alongside to the runway 28 approach lights. Even-numbered rooms in 222-260 range look out on to the approach and you can take decent photographs of arrivals. It's a short walk to the Old Airport Rd spotting area, and there's a free shuttle to the terminal.

Public Transport: Bus routes #16 and #41 link Dublin Airport with the city centre.

Dublin Weston

Runways

07/25 3,030ft / 1,381m

Weston is Dublin's second airport, located 9 miles west of the city. It is primarily a general aviation airport, but sees executive aircraft movements regularly. You can see most parked aircraft from the car park outside the terminal building.

Public Transport: Bus routes #C1, #C2 and #52 from central Dublin passes close to Weston. Alight at Cooldrinagh Lane and walk the last 15 minutes to the terminal.

Ireland West Airport

Runway

08/26 7,700ft / 2,340m

Located at Knock in County Mayo, Ireland West Airport is the country's fourth busiest, with flights by Aer Lingus Regional and Ryanair.

The airport has a single runway, with a small passenger terminal and main parking apron in the north east corner. A smaller apron is at the centre point of the runway, to its north, where in recent years a former Air France Airbus A380 arrived for part-out and scrapping, along with some other airliners.

When approaching the airport along the main R376 access road, off the N17, turn left along a track which disappears up a hill (opposite a lay-by) before you reach the terminal. This track becomes unpaved and rough, but if you follow it you'll find various vantage points on both the north and south side of the runway at the 26 end. These are sufficient for photographing aircraft on approach and on the ground.

Another road runs from the terminal car park towards the small business park and area where airliner scrapping takes place.

Kerry

Runways

08/26 6,562ft / 2,000m

Kerry is a small airport in western Ireland served by Ryanair on a handful of domestic and European routes. It is very quiet, with little general aviation or other services.

Any aircraft parked outside the terminal can be seen from the car park, but those at the various hangars are difficult to see unless they taxi out to the runway.

Public Transport: Buses link Kerry Airport with Tralee, Killarney and Limerick. There is a railway station in Tralee.

Shannon

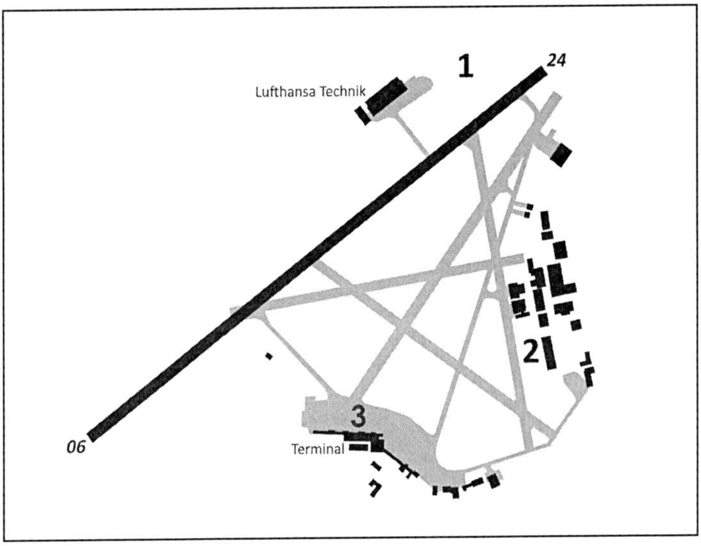

Runways

06/24 10,495ft / 3,199m

Hub For: Ryanair

No longer a regular transatlantic stop-off point for many airliners, Shannon still sees its fair share of traffic crossing the Pond. Seasonal links with United Airlines still exist, and Aer Lingus links the airport with Boston and New York year-round. Otherwise, traffic consists of summer charters, and European schedules Ryanair in particular. Cargo traffic is fairly busy with carriers such as DHL, FedEx Express, Turkish Cargo and UPS.

The passenger terminal occupies the southern side of the airport. Maintenance and storage areas are on the eastern and northern peripheries of the airport, with a main runway running south west to north east.

Often, you'll see exotic aircraft at Shannon undergoing maintenance or painting, or in storage.

SPOTTING LOCATIONS

1. Runway 24 Viewing Area

There is a public viewing area next to the large hangar on the northern perimeter of the airport, close to the runway 24 threshold. The majority of aircraft use this direction for landing, so you will see most movements. The fence is a little high for photographs unless you have a step ladder. From the terminal, head towards Galway and Ennis on the N19. Follow signs for R472 Newmarket-on-Fergus, then L3169 Ballycally. When you see the hangar, turn left, then left again.

2. Eastern Aproans

Exploring the Shannon Free Zone industrial estate on the eastern side of the airport should give you views of some of the aircraft parked outside the maintenance hangars. From the terminal, head towards Galway and Ennis on the N19. Turn left at the roundabout into the Free Zone and follow Lindbergh Avenue in particular.

3. Aviation Gallery

A viewing area exists inside the terminal, for departing passengers only. You can see the apron through windows, and also enjoy the large collection of diecast aircraft models.

Public Transport: Buses #51 and #343 link Shannon Airport with central Limerick.

Waterford

Runways

03/21 4,700ft / 1,433m

A small regional airport which at present has no scheduled services. The only operations here are light aircraft and ad-hoc charters. The car park outside the terminal is usually sufficient to see any aircraft parked up.

Republic of Ireland's Military Airfields

Casement Aerodrome

Runways

04/22 4,800ft / 1,463m
10/28 6,001ft / 1,829m

Ireland's premier military airfield is Baldonnel, or Casement Aerodrome, located outside Dublin.

It is home to the Irish Air Corps, as well as government flying, and often receives foreign dignitaries.

The airfield has two runways, with a range of hangars and a main apron on the northern side, off Baldonnel Road.

Base visits are sometimes possible when arranged in advance. Otherwise it is difficult to spot from anywhere outside the base, other than seeing fleeting views of aircraft arriving or departing.

Aviation Museums & Collections

There are many great aviation museums and collections scattered around the United Kingdom and Ireland. Here are some of the favourites and most substantial:

Aerospace Bristol

Hayes Way, Patchway, Bristol BS34 5BZ | www.aerospacebristol.org

One of the newest aviation museums in Britain. Aerospace Bristol is located on the former Filton Airfield site to exhibit the history of aircraft manufacturing in the city. It includes the last Concorde to fly and other aircraft.

Open daily except Monday 10am-4.30pm (open Mondays in school summer holidays). Adults £18, Concessions £16, Children £10, Family £25/£42/£52, Under 4s & Essential carers Free.

AeroVenture South Yorkshire Aircraft Museum

Dakota Way, Airbourne Road, Doncaster, DN4 7NW | 01302 761616 | *www.southyorkshireaircraftmuseum.org.uk*

A large collection of early aircraft, fast jets, helicopters and smaller civil airliners. Most inside large hangar, with some outside and in smaller buildings.

Open Tuesday-Sunday 10am-5pm (late March-late October), 10am-4pm (November-March). Adults £7, Seniors £6, Children £3.50, Family £16.50. Under 5s free.

Bournemouth Aviation Museum

Merritown Lane, Hurn, Christchurch, BH23 6BA | 01202 473141 | *www.aviation-museum.co.uk*

A compact collection of vintage jets, warbirds, airliners and helicopters, mostly just cockpit sections but some complete. Located next to Bournemouth Airport.

Open daily 10am-5pm (late March-late October), 10am-4pm (late October-late March). Adults £7, Children £3, Under 3s free.

Brooklands Museum

Brooklands Road, Weybridge, Surrey, KT13 0QN | www.brooklandsmuseum.com

Occupying the historic Brooklands race circuit and former aircraft factory where many British types were built. Museum combines aircraft and automotive history. Impressive collection of airliners (VC-10s, Vanguard, Varsity, Viscounts, Concorde prototype) and wartime aircraft. Many are open to enter.

Open daily 10am-5pm (March-late October), 10am-4pm (November-February). Adults £18.10, Seniors £17.20, Children £10, Family £29.60/£47.70, Under 5s free.

City of Norwich Aviation Museum

Old Norwich Road, Horsham St Faith, Norfolk, NR10 3JF | 01603 893080 | *www.cnam.org.uk*

Small museum on the north side of Norwich Airport. Has complete Nimrod, Vulcan, HP Herald, Fokker F-27, Lightning, Jaguar, Hunter, Meteor and other types.

Open Tuesday-Saturday 10am-5pm, Sunday & public holidays 12-5pm (April-October); Wednesday & Saturday 10am-4pm, Sunday 11am-3pm (November-March). Adults £9, Concession £7, Children £5, Family £25.

Cornwall Aviation Heritage Centre

HAS 3, Aerohub 2, St Mawgan, Newquay, Cornwall, TR8 4JN | 01637 861962 | *www.cornwallaviationhc.co.uk*

A fantastic collection of complete aircraft, active in many cases. They include a Canberra, Sea Devon, VC-10, Hunters, Harrier, BAC One-Eleven and Varsity.

Open daily except Friday & Saturday, 10am-4pm. Adults £10.50, Concessions £8, Children (under 14) £6, Family £26.

de Havilland Aircraft Museum

Salisbury Hall, London Colney, Hertfordshire, AL2 1BU | www.dehavillandmuseum.co.uk

Celebrating the history of this famous British aircraft designer and manufacturer. Exhibits include most de Havilland types, plus some of the larger designs including Comet 1, BAe 146 and Trident.

Open Tuesday to Sunday, plus public holidays, 10.30am-5pm. Adults £12, Concessions £10.90, Children £8, Family £35. Under 5's and carers of disabled free.

Dumfries & Galloway Aviation Museum

Heathlands Industrial Estate, Dumfries, DG1 3PH | 01387 251623 | *www.dumfriesaviationmuseum.com*

On the site of a former airfield. Collection is mostly military aircraft and helicopters, including Lightning, Canberra, Hunter, F-100 Sabre, Fairey Gannet. Also has forward section of Trident 3B and a Jetstream.

Open Wednesday, Saturday & Sunday 11am-5pm. Adults £7, Senior/Children £5, Family £20.

Duxford Imperial War Museum

Duxford, Cambridgeshire, CB22 4QR | *www.iwm.org.uk*

A historic airfield and one of the country's best aircraft collections. Includes many historic warplanes, fighters, airliners, and an on-site American Air Museum with its own interesting aircraft. Airliners include Concorde, VC-10, Trident 2E, BAC One Eleven, Viscount and Bristol Britannia.

Open daily 10am-6pm. Adults £25, Concessions £22.50, Children (5-15) £12.50, Under 5s and IWM Members free.

East Midlands Aeropark

Castle Donington, Leicestershire, DE74, 2PR | www.eastmidlandsaeropark.org

A small outdoor collection of historic aircraft including Canberra, Hunters, Nimrod, Buccaneer, Lightning, Vulcan, Argosy, Dove, Varsity, plus sections of Viscount, VC-10 and Vanguard.

Open Saturday, Sunday & Thursday, 11am-5pm (April to late October). Closed November-April.

Adults £5, Children £3, Under 5s free.

FAST Museum

85 Farnborough Road, Farnborough, Hampshire, GU14 6TF | 01252 375050 | *www.airsciences.org.uk*

Dedicated to the Farnborough's contribution to air sciences and aviation history. Many exhibits, with a small collection of jets and helicopters outside.

Open Saturday, Sunday and public holidays, 10am-4pm. Free entry.

Fleet Air Arm Museum

RNAS Yeovilton, BA22 8HW | 01935 840565 | *www.fleetairarm.com*

Many aircraft and artefacts relating mostly to the Royal Navy and its history. Also houses one of the Concorde prototypes.

Open Wednesday-Sunday, 10am-4.30pm. Adults £19, Concessions £18, Children £14, Family £39/£49, Under 3s free.

Jet Age Museum

Meteor Business Park, Cheltenham Road East, Gloucester, GL2 9QL | *www.jetagemuseum.org*

A small museum at Gloucestershire Airport displaying mostly Gloster aircraft such as the Javelin and Meteor, plus sections of a Vulcan, Trident 3B and Typhoon.

Open Saturday & Sunday, 10am-4pm.

Lincolnshire Aviation Heritage Centre

East Kirkby, Lincolnshire, PE23 4DE | 01790 763207 | *www.lincsaviation.co.uk*

Home to one of two operable Avro Lancasters in Britain. Does taxi runs with you on board! Also historic World War II base.

Open Tuesday-Saturday 9.30am-5pm (Easter-late October), 10am-4pm (late October-Easter). Adults £10, Seniors £9, Children £4, Family £26, Under 6's free.

Manchester Runway Visitor Park

Sunbank Lane, Altrincham, WA15 8XQ | *www.runwayvisitorpark.co.uk*

Widely used as a spotting location at Manchester Airport, the Runway Visitor Park is also home to a collection of aircraft including Avro RJX, Concorde, Trident 3B, Nimrod, and McDonnell Douglas DC-10 forward fuselage. Most are open to the public.

Open daily except 25-26 December from 8am. Closes 4pm (November-February), 5pm (March), 6pm (April-May/September-October), 8pm (June-August). Free entry but car parking charges apply (£5-£12).

Midland Air Museum

Coventry Airport, Bagington, Warwickshire, CV3 4FR | *www.midlandairmuseum.co.uk*

Nice outdoor collection include a Vulcan, Argosy, Viscount, Dove, Meter and many more aircraft from across the spectrum.

Open daily except 24-26 December, 10am-4.30pm (November-March), 10am-5pm (April-October). Adults £8.25, Concessions £7.75, Children £4.25, Family £23.50, Infants free.

National Museum of Flight

East Fortune Airfield, East Lothian, EH39 5LF, Scotland | 0300 123 6789 | *www.nms.ac.uk/national-museum-of-flight*

A significant aviation museum with many indoor and outdoor aircraft exhibits covering British aviation history in particular.

Airliners on display include a Comet 4, BAC One-Eleven, Concorde, Dove and Twin Pioneer, plus forward sections of Boeing 707 and Trident 1C. Also many military and wartime aircraft.

Open daily 10am-5pm (late March-late October), Saturday-Sunday 10am-4pm (November-March). Adults £12.50, Concessions £10.50, Children £7.50, Family £33, Under 5s and NMS members free.

Newark Air Museum

Drove Lane, Newark, Nottinghamshire, NG24 2NY | 01636 707170 | *www.newarkairmuseum.org*

Large collection of significant aircraft, both civil and military, from early conflicts to recent times. Nicely presented in various hangars and outdoor areas.

Open daily except 24-26 & 31 December, 10am-5pm (March-October), 10am-4pm (November-February). Adults £10, Seniors £9, Children £5, Family £27, Under 5's free.

North East Land, Sea and Air Museum (NELSAM)

Old Washington Road, Sunderland, Tyne & Wear SR5 3HZ | *www.nelsam.org.uk*

Hangar full of exhibits, mostly fighter jets and helicopters. Outside a Vulcan, Trident 1C, Lightning and Canberra. Also includes military vehicles, buses and trams.

Open Tuesday-Sunday 10am-4pm (except 25, 26 and 31 December). Adults £6.50, Concessions & Children £3.50, Family £18, Under 5s free.

Royal Air Force Museum Cosford

Shifnal, Shropshire, TF11 8UP | 01902 376 200 | *www.rafmuseum.org.uk/cosford*

A large museum with many interesting exhibits related to the history of the Royal Air Force. Mostly military aircraft, but also includes complete DC-3, Viscount, Bristol Britannia and Comet 1.

Open daily 10am-5pm (March-October), 10am-4pm (November-February). Free entry.

Royal Air Force Museum Hendon

Grahame Park Way, London, NW9 5LL | 020 8205 2266 | *www.rafmuseum.org.uk/london*

The other museum telling the history of the Royal Air Force with some significant aircraft and designs on display on the historic Hendon Airfield site. Includes Lancaster, Vulcan, B17G Flying Fortress, Spitfires, Hurricanes and many World War I aircraft.

Open daily 10am-5pm. Free entry.

Science Museum, London

Exhibition Road, South Kensington, London, SW7 2DD | *www.sciencemuseum.org.uk*

Large museum in central London. The aviation hall has a number of complete aircraft and sections to explain the history and science of flight.

Open daily, 10am-6pm. Free entry.

Science Museum, Wroughton

Red Barn, Wroughton, Swindon, Wilstshire, SN4 9LT | *www.sciencemuseum.org.uk*

Acting as a store for the Science Museum's largest collections, it comprises many complete aircraft and airliners, including Lockheed Constellation, Boeing 247, Douglas DC-3, Trident 3B, Comet 4B.

Currently closed. Expected to reopen 2024.

Shannon Aviation Museum

Link Road, Shannon, Co. Clare, Rep Ireland | +353 61 363687 | *www.shannonaviationmuseum.com*

An interesting museum in Shannon, Co Clare, with a selection of smaller aircraft types from gliders to light aircraft and wartime props and jets. Also includes a selection of cockpits, a BAC 1-11 forward fuselage and a 737 simulator.

Open Wednesday-Saturday (summer) 10am-4pm. Adults €12, Concessions €9, Children 0-12 €6.

Solway Aviation Museum

Carlisle Airport, Cumbria, CA6 4NW | 01228 573823 | *www.solway-aviation-museum.co.uk*

Mostly military collection, with complete Vulcan, Canberra, Hunter, Lightning, Phantom. Trident 1C cockpit.

Open: Fri, Sat, Sun and public holidays from 1 April-30 October. Adults £6, Senior/Child £4, Family £15.

The Helicopter Museum

Locking Moor Road, Weston-Super-Mare, BS24 8PP | 01934 635227 | *www.helicoptermuseum.co.uk*

The largest dedicated helicopter museum in the world. Displays around 100 examples across the spectrum from around the world.

Open daily, 10am-5.30pm (April-October), 10am-4.30pm (November-March). Adults £12, Senior £10, Children £6, Family £30/£32.

South Wales Aviation Museum

Picketston Business Park, CF62 4WB | 01446 788476 | *www.swam.online*

A great little aviation museum just to the north of MOD St Athan in South Wales. It has a collection of civilian and military jets inside a large hangar, including a number of cockpits. Recent additions include a former Queen's Flight BAe 146. A number of aircraft sections and parts can be seen in the yard behind the hangar.

Open Saturday, Sunday and Bank Holidays, 10am-4pm. Adults £7.50, Concessions/Veterans £5, 1 child free.

The Shuttleworth Collection

Old Warden Aerodrome, Bedfordshire, SG18 9EP | 01767 627 927 | *www.shuttleworth.org*

Britain's premier collection of flyable vintage aeroplanes, most from World War I era. The collection often takes part in air displays particularly at its Old Warden home. When not flying visitors can see them at close quarters.

Open daily 10am-5pm (February-October), 10am-4pm (November-February). Check event dates on website. Adults £19.50, Children & Essential carers free.

Ulster Aviation Society

Gate 3 — Maze Long Kesh, 94-b Halftown Road, Lisburn, BT27 5RF, Northern Ireland | www.ulsteraviationsociety.org

The premier aviation museum in Northern Ireland, situated at the former RAF Lisburn, south west of Belfast. The collection includes fast jets, helicopters, trainers, wartime aircraft, and aircraft produced locally by Shorts Brothers.

Visits by prior arrangement only (see website). Up to 4 slots available daily.

Yorkshire Air Museum

Elvington, York, YO41 4AU | 01904 608595 | *www.yorkshireairmuseum.org*

Occupying the wartime part of Elvington Aerodrome, with many of the historic buildings preserved. Aircraft occupy various hangars, ranging from World War I types to Cold War jets. Highlights include Douglas DC-3, Avro Victor, Nimrod, Halifax and Mosquito. Some aircraft are taxied on occasion.

Open daily except 25-26 December and 1 January, 10am-4pm (Winter), 10am-5pm (Summer). Adults £12, Children £5, Students £9, Families £30, Under 4's & Military veterans free.